Cathodic Protection Survey Procedures

AF779041

Cathodic Protection Survey Procedures

W. Brian Holtsbaum

NACE INTERNATIONAL
1440 South Creek Drive
Houston, Texas 77084

NACE INTERNATIONAL

The Corrosion Society

© 2009 by NACE International
All rights reserved

Printed in the United States of America

Library of Congress Cataloguing-in-Publication Data

Holtsbaum, W. Brian.
Cathodic protection survey procedures / W. Brian Holtsbaum.
p. cm.
Includes bibliographical references and index.
ISBN 978-1-57590-219-7 (pbk.)
1. Cathodic protection. I. Title.
TA418.74.H65 2009
620.1'1223–dc22 2008042239

ISBN 978-1-57590-219-7

Neither NACE International, its officers, directors, or members thereof accept any responsibility for the use of the methods and materials discussed herein. The information is advisory only and the use of the materials and methods is solely at the risk of the user.

This book, or any parts thereof, may not be reproduced in any form without permission of the copyright owners.

NACE International
1440 South Creek Drive
Houston, Texas 77084
http://www.nace.org

Contents

Preface

When I started in cathodic protection (CP) in 1957, there was very limited information on CP testing and design, but I had the good fortune to have Marshall Parker as my mentor. He prepared one of the first CP books on field tests, although admittedly, one had to be experienced to understand the implications of several of the statements in his book. Having received instruction from the author, that book was very useful. A.W. Peabody published the next very helpful text, *Control of Pipeline Corrosion*, and the second edition of that book is still widely used today. Even with this information, many procedures had to be developed to complete field tests and to design, commission, and troubleshoot various CP systems.

Over the following years, when training new CP recruits, I often wished that there was a book covering detailed, step-by-step testing procedures for them to study and have as a reference. Those who are delegated to training others often have many additional pressing duties, so their trainees have to supplement their knowledge from other sources. NACE is a huge resource for our industry, and it is hoped that this book will be a useful addition to its resources.

Several texts and many papers that describe the theory and application of CP are referenced in this book. This book does not revisit those areas because it is assumed that the field person will have this background knowledge and that he or she is accustomed to the safe use of instruments. Instead, this manual is intended to provide a step-by-step survey procedure for tests related to CP. The book is divided into stand-alone procedures and, as a result, some repetition does exist among these procedures. Even so, some cross referencing between procedures is necessary.

Training as an NACE CP Tester or an NACE CP Technician, or the equivalent, is recommended at the minimum to complete these field procedures. The analysis of work should be completed by persons who have the equivalent of an NACE CP Technologist or NACE CP Specialist certification, with appropriate recognition of local jurisdictions.

I would like to extend my gratitude to Kevin Garrity, who reviewed my manuscript.

Finally, I offer my encouragement to the CP Testers, Technicians, Technologists, and Specialists of the world to gather and accept only accurate, legible, and well-documented data to help ensure the preservation of our future infrastructure and the safety of the public.

W. Brian Holtsbaum, P. Eng.
NACE Corrosion Specialist

CATHODIC PROTECTION PROCEDURE NO. 1

Rectifier Adjustment, Inspection, and Basic Troubleshooting 1

1.0 INTRODUCTION

The purpose of this procedure is to outline a procedure to adjust, inspect, and perform basic troubleshooting tasks on a cathodic protection rectifier. A person must be trained and qualified to perform each specific task safely.

The basic components of a cathodic protection rectifier normally consist of the following:

- A rectifier case and panel
- Alternating current (AC) voltage input terminals (and taps if a dual-voltage input)
- AC circuit breaker mounted on the panel
- Transformer (and taps if a constant voltage rectifier)
- Rectifying elements
- Panel voltmeter and ammeter c/w shunt*
- Direct current (DC) output terminals
- AC/DC fuses*
- AC and DC surge protection*

Rectifiers for cathodic protection service can be categorized as constant voltage, constant current, and constant potential. The majority of the rectifiers in use are a constant voltage type, and as a result, most of the discussion

* May be omitted in some cases.

Table 1.1 Suggested NACE Qualifications for Specific Activities

	Qualification for Activity			
CP Procedure No. 1	CP Tester or Equivalent	CP Technician, Corrosion Technician with CP, or Equivalent	CP Technologist, Corrosion Technologist with CP, or Equivalent	CP Specialist, Corrosion Specialist with CP, or Equivalent
Section 1.0	X	X	X	X
Section 2.0	X	X	X	X
Section 3.0	X	X	X	X
Section 4.0	X	X	X	X
Section 5.0	X	X	X	X
Section 6.0				
6.1.1–6.1.6	X	X	X	X
6.1.7, 6.1.8		X	X	X
6.1.9–6.1.16	X	X	X	X
6.1.17, 6.1.18		X	X	X
6.1.19			X	X
6.2	X	X	X	X
6.3.1	X	X	X	X
6.3.2		X	X	X
6.3.3	X	X	X	X
6.4	X	X	X	X
6.5	X	X	X	X
Section 7.0				
7.1	X	X	X	X
7.2	X	X	X	X
7.3.1–7.3.5	X	X	X	X
7.3.6–7.3.9		X	X	X
7.3.10	X	X	X	X
Section 8.0				
8.1		X	X	X
8.2		X	X	X
8.3	X	X	X	X
8.4		X	X	X
8.5	X	X	X	X
8.6		X	X	X
8.7.1	X	X	X	X
8.7.2		X	X	X
8.7.3–8.7.5	X	X	X	X
8.8	X	X	X	X
8.9	X	X	X	X
8.10		X	X	X
8.11	X	X	X	X
8.12	X	X	X	X

is related to this type of rectifier. The differences between constant voltage rectifiers and the other types of rectifiers are discussed.

The analysis of data recorded is given in each section, as appropriate.

Rectifier readings, adjustment, inspection, and troubleshooting rectifiers are often completed while working alone, and safety is paramount. A company's working alone policy must be met in these circumstances.

Although possible hazards are covered in the following sections, they cannot be expected to cover all of the hazards that one might encounter. Most companies have a process to identify hazards, determine the level of risk, and develop preventative measures for each task or project. Such a process should be followed before starting work. If the hazards change during the work, a new assessment should be completed and appropriate changes made. An example of a change in hazard is for a person to expand testing on the panel to testing behind the panel with the AC power still energized. Testing of components under live loads by qualified persons may be needed for troubleshooting. Measures that may be required could include the following:

- An increase in the qualification of the person doing the work
- A change in the work permit, as necessary
- If applicable, readdressing your working alone procedure for the job and perhaps making more frequent contact with the safety contact when the risk increases

Table 1.1 relates the suggested NACE certification for a person assigned to an activity, as described within this procedure. Certainly, those with other applicable training, such as electricians, may also be qualified.

Be alert and safe and, if hazards are encountered that are not highlighted in this procedure, formally alert others to the hazard and take steps to reduce the hazard.

2.0 TOOLS AND EQUIPMENT

Select tools and equipment that are appropriate for work on electrical components and offer the level of personal protection for the hazard involved. The following tools are the minimum that will be required:

- Multimeter capable of measuring 1 mV_{DC} to 100 V_{DC} and 250 V_{AC} (or exceeding the AC input voltage), complete with leads with insulated probes; a higher AC voltage may be necessary, depending on the AC supply voltage

- Optional clamp-on AC/DC ammeter sized for AC and DC current rating of the rectifier
- Tools, including:
  - Wrench/socket drive to fit the adjustment taps, if applicable
  - Screwdriver of the type and size for terminal strip type of tap
  - Small screwdriver sized to adjust circuit board potentiometers, if applicable
- Temporary resistance load sized for the rated current and wattage of the rectifier, if the rectifier is to be run on a temporary load
- Wrench to fit DC terminals

3.0 SAFETY EQUIPMENT

- Standard safety equipment and clothing, as required by the company's safety manual and regulations
- Electrical lock-out/tag-out equipment
- Electrically insulated clips and probe handles for meter leads
- External electrical disconnect to the rectifier
- An instrument that alerts one to high voltage without touching

4.0 PRECAUTIONS

4.1 *Caution!* Exposed bare terminals may be at a high voltage. Wherever possible, select the test that can be performed with the rectifier turned off and locked out/tagged out.

4.2 The person working on a rectifier must be qualified and/or hold proper certification in accordance with local regulations and company policy to conduct tests on and adjustments to the rectifiers.

4.3 Prior to touching a rectifier case, take a case to ground AC voltage measurement to confirm that it is safe. Alternately, an instrument is available that, when held near the rectifier, will turn on a light if a hazardous voltage exists. Never grab a case latch or lock with your hand because a dangerous voltage will cause the fingers to contract, preventing you from letting go. Touching the case with the back of the hand has been recommended in the past because a contraction in the fingers will not lock your hand onto the case. This still allows a shock and is no longer considered a safe practice.

4.4 The rectifier must be turned off for all rectifier adjustments. Turn off the external AC disconnect prior to working inside the rectifier. Lock-out/tag-out equipment must be in place before working on rectifier components.

4.5 An inoperative rectifier is to be initially energized at a low tap setting.

4.6 Open the rectifier case very carefully as it may have become a residence for dangerous insects (bees, wasps, or hornets), snakes, or rodents. Entry holes should be blocked in this event.

4.7 For air-cooled rectifiers, ensure that all screened areas are clean and that there is proper air circulation throughout the rectifier components. Any holes in the screens should be repaired or the screen replaced.

4.8 For oil-cooled rectifiers, ensure that the oil is clear and to the correct level before energizing. If an older oil-cooled rectifier, confirm that the oil does not contain polychlorinated biphenyls. If it does, do not contact the oil and wear protective clothing.

4.9 Review the monitoring data to determine the rectifier operating history.

4.10 Before adjusting a rectifier with an abnormal DC voltage or DC current output, complete an inspection of the rectifier or conduct a diagnostic investigation of the structure network, including bonds, to determine the cause.

4.11 Adjusting a rectifier without knowing the cause of the change in output may result in more damage to the structure. For example, if a structure bond is broken, increasing the current may increase stray current interference in that area.

5.0 RECTIFIER COMPONENTS

5.1 General

A general understanding of the components and their function in a cathodic protection rectifier must be known.[9.1] Although all options may not be installed in each rectifier, the order of the components for a constant voltage rectifier is as follows:

1. AC supply
2. Primary transformer taps (if a dual-AC-voltage input)
3. AC surge protection
4. AC circuit breaker
5. Transformer
6. Secondary taps
7. AC fuse
8. Rectifying elements
9. DC fuse
10. Shunt/ammeter and voltmeter
11. DC surge protection
12. DC output terminals
13. Noise/efficiency filters (optional)

5.2 Alternating Current Supply and Primary Alternating Current Transformer Taps

5.2.1 The AC supply is normally brought into terminal blocks that are located at the back or bottom of the rectifier.

5.2.2 If the rectifier is a dual-AC-voltage input, the primary of the transformer is divided into two parts, which can either be placed in series for the higher voltage input or in parallel for the lower voltage output. Link bars or wires are located near the AC supply with directions for their position for either the lower or higher AC voltage input. Note that the rectifier can operate when set for the higher AC voltage but supplied with a lower AC voltage, except that only one-half the rated DC voltage output can be obtained. Damage to the transformer can be expected if set for the lower AC voltage input and supplied with the higher AC voltage.

5.3 Alternating Current Surge Protection and Alternating Current Breaker

5.3.1 The circuit breaker is a mechanical switch that has either a thermal or an electromagnetic trip element that activates with a current beyond the rating of the breaker. Most rectifiers now have a magnetic type of breaker.

5.3.2 Turn the circuit breaker on in two stages: first, hold it so that it makes contact; then, push it further until it locks in place. Attempting to turn it on quickly and lock it at the same time may cause it to trip.

5.3.3 AC surge protection may be located across the breaker or across the AC supply. This surge protection is an attempt to prevent power line faults from going to ground through the rectifier and, for this reason, it often fails as a short, causing the AC breaker to trip.

5.4 Transformer and Secondary Taps

5.4.1 The transformer consists of primary and secondary windings that are coiled around a magnetic steel core. The ratio of the number of windings in the primary versus the secondary will determine the change in AC voltage from the AC input into the primary windings to the AC voltage output from the secondary windings. Again, compare the AC voltage supply with that rated on the nameplate to confirm that the correct AC voltage is being supplied.

5.4.2 In a constant voltage rectifier, the secondary windings are tapped to allow a selection of AC voltage into the rectifying element (Figure 5.1). Coarse taps are attached at a uniform spacing along the secondary windings. Within the coarse tap at one end of the secondary winding, fine taps are attached to provide better control. The sum of the fine tap voltages will, therefore, approximate that of a coarse tap voltage.

5.4.3 A three-phase rectifier will have three sets of coarse and fine taps, one for each phase. These must be set at the same coarse and fine settings to balance the phases.

5.5 Alternating Current Fuse

5.5.1 A fuse is a low melting point, metallic element that protects the more expensive components.

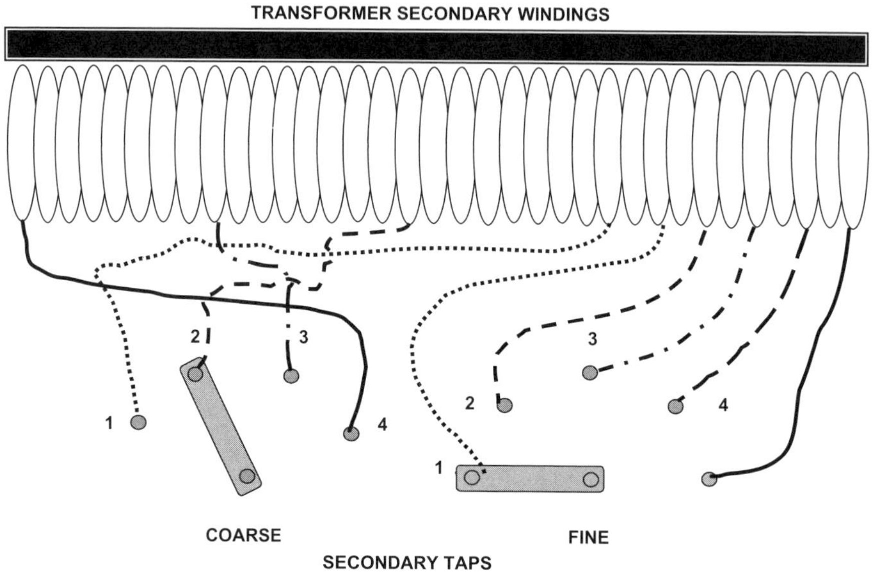

Figure 5.1 Typical secondary transformer taps for a constant voltage rectifier are shown.

5.5.2 An AC fuse between the transformer and the rectifying elements is optional and is sized to protect the rectifier elements and the transformer secondary windings.

5.5.3 There should be no voltage across a fuse. A voltage drop across a fuse indicates that it has blown.

5.5.4 After de-energizing the rectifier, remove and test the fuse for continuity.

5.6 Rectifying Elements

5.6.1 The role of the rectifying element is to convert AC to DC.

5.6.2 The rectifying element has the characteristic that it will allow current to pass in one direction but will block it in the opposite direction (polarity). The most common materials in cathodic protection rectifiers are selenium plates or silicon diodes.

5.6.3 A single element will allow one-half of the AC cycle to pass but will block the reverse cycle, thus, providing a one-half-wave DC output with a high ripple. A second element can be installed in the circuit to allow the reverse cycle to pass yet go in the same

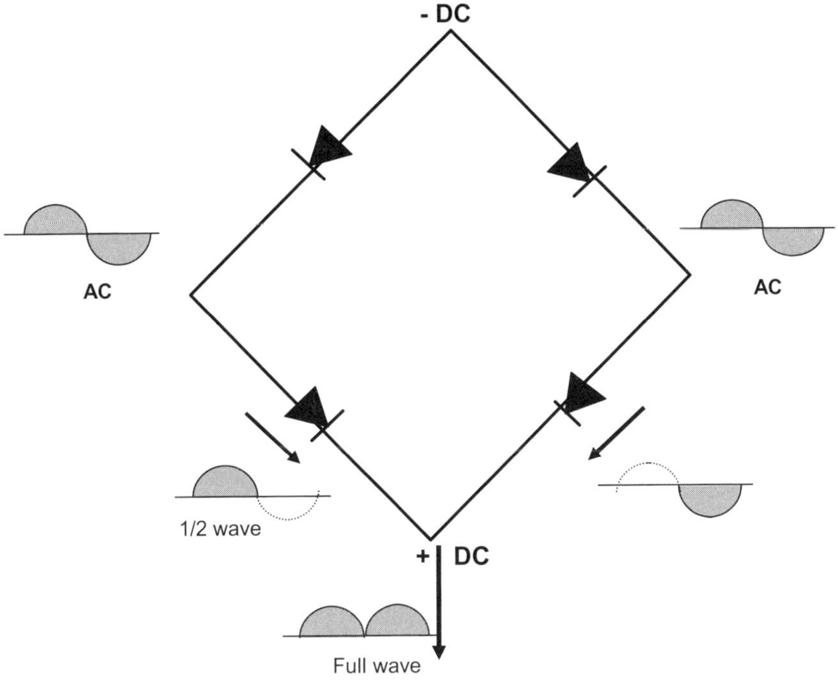

Figure 5.2 Rectifier bridge circuit is shown.

direction for the reverse cycle. Such a circuit, which provides a full-wave output, is shown in Figure 5.2.

5.6.4 Selenium plates are installed in series for voltages over 20 V due to their low barrier voltage of 25–30 V. Silicon diodes are sized for the rated output.

5.7 Direct Current Fuse

5.7.1 A DC fuse between the rectifying elements and the DC output terminals is intended to protect these elements from a surge, likely coming from the pipeline.

5.7.2 A DC fuse will not be found in all rectifiers.

5.8 Meters and Shunt

5.8.1 Panel-mounted DC voltmeters and ammeters are installed in most rectifiers to monitor their operation.

5.8.2 The DC voltmeter records the voltage across the DC output terminals.

5.8.3 The DC ammeter may have an internal shunt, but in most cases, the shunt is mounted on the rectifier panel. The ammeter must be calibrated for the shunt rating.

5.8.4 Panel meters are often found out of calibration, due, in part, to the environment to which they are exposed. They should only be used as an indicator of the rectifier operation and the actual output, confirmed with a portable meter.

5.9 Direct Current Output and Direct Current Surge Protection

5.9.1 The DC output terminals are labeled positive (+) and negative (−) and may also have an anode label by the positive terminal and a structure label at the negative terminal. It is imperative that the *positive* terminal be connected to the *anodes* and the *negative* terminal to the *structure*; otherwise, corrosion will be accelerated on the structure. The polarity must be confirmed by a portable meter.

5.9.2 DC surge protection is often provided to allow surges coming from the structure to go to ground via the anodes, without passing through and damaging the rectifying elements.

6.0 MEASUREMENT PROCEDURE

- Complete only the tests for which you are qualified and/or hold certification consistent with local regulations (see also Table 1.1). Otherwise, get help from someone who is qualified and/or holds the proper certification.
- Turning the rectifier power (AC disconnect) off includes using a lock-out/tag-out procedure.
- *Important!* Measure the ON/OFF structure-to-electrolyte potential after the rectifier is energized to ensure that it becomes more electronegative in value when the rectifier is turned on. This will confirm that the rectifier positive is connected to the anodes and the negative to the structure. Reversing this connection will cause accelerated corrosion damage to the structure.

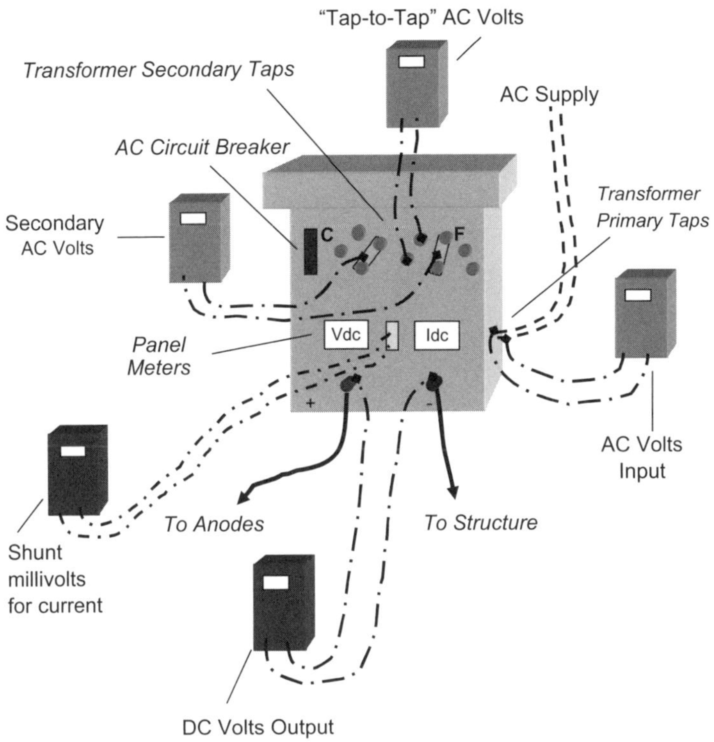

Figure 6.1 Initial rectifier voltage measurements are shown.

6.1 Energizing New or Re-Energizing a Tap-Type, Constant Voltage-Type Rectifier

6.1.1 Measure a structure-to-electrolyte potential near the rectifier before it is energized.[1] Repeat this measurement after it is energized.

6.1.2 Ensure that the AC power disconnect outside the rectifier is locked out until ready to energize.

6.1.3 If the rectifier is of a dual AC voltage input type, inspect the AC input link bars or bonding straps to ensure that they match the AC voltage supply (see Figure 6.1). If adjustment is required, ensure that the outside AC disconnect is off.

---

1 *See* Cathodic Protection Procedure No. 2: Structure-to-Electrolyte Potential Measurement.

6.1.4 Measure the DC voltage between the structure and the anode before energizing. There is often a back electromotive force (EMF) that opposes the rectifier DC voltage (see Figure 6.1) due to the galvanic potential difference between the structure metal and the anode material or the carbon in the coke breeze backfill.

6.1.5 Disconnect the DC load (positive or negative cable). The back EMF voltage will drop to 0 V. Leave the cables disconnected.

6.1.6 Ensure that the rectifier AC circuit breaker between the AC supply and the transformer is off.

6.1.7 Measure the primary AC supply voltage (see Figure 6.1).

6.1.8 Turn the rectifier AC circuit breaker on.

6.1.9 Measure the tap-to-tap AC voltage between transformer secondary taps (see Figure 6.1). The voltage between the coarse taps should be approximately equal. The voltage between the fine taps will also be approximately equal, and their sum should equal that of one of the coarse taps. A tap-to-tap voltage of 0 V_{AC} indicates a faulty secondary tap wire or open windings.

6.1.10 Turn the rectifier AC circuit breaker off and reconnect the DC cable(s).

6.1.11 Set the AC secondary taps to a low tap setting but to a DC voltage greater than the DC voltage measured in 6.1.4.

6.1.12 The rectifier output DC voltage is controlled by adjusting the secondary AC taps using the following procedure:

6.1.12.1 Set the coarse ("C") tap to the lowest labeled number or letter ("1" or "A").

6.1.12.2 Set the fine ("F") tap to the lowest setting ("1", normally).

6.1.12.3 Energize the rectifier and record the DC volts and DC amperes.

6.1.12.4 Note that the rectifier will not impress any current until the open circuit back EMF between the anode and the structure, as measured in 6.1.2, has been exceeded.

6.1.12.5 Turn the rectifier off before adjusting the taps each time.

6.1.12.6 Increase the voltage by increasing the fine tap to 2 (F-2, etc.) until either the required current has been reached or there are no more fine taps (Figure 6.2).

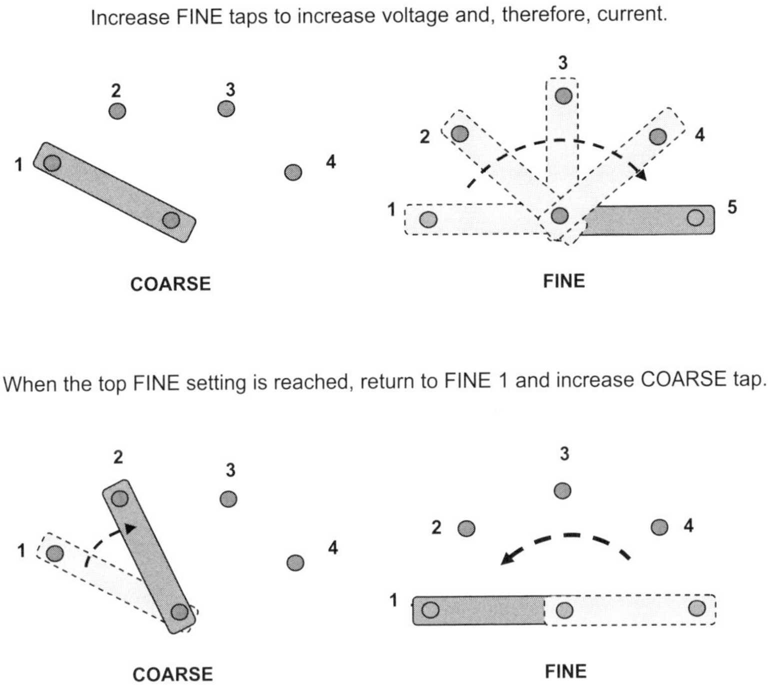

Figure 6.2 Procedure to change rectifier taps to increase secondary AC voltage and, thus, current, is shown.

6.1.12.7 If more current is still needed, reconnect the fine tap to 1 and move the coarse tap to 2 (C-2).

6.1.12.8 Increase the fine taps until either the required current or the maximum fine tap is reached.

6.1.12.9 If more current is needed, return the fine tap to 1 (F-1) and move the coarse tap to 3 (C-3).

6.1.12.10 Repeat this procedure until the required current is obtained.

6.1.12.11 If the tap settings need to be reduced, the procedure is reversed by moving the fine taps to a lower number. When the minimum fine tap is reached, reduce the coarse tap to the next lower number or letter and increase the fine taps, as necessary.

6.1.13 If working on a three-phase (3Φ) AC voltage rectifier, set the three coarse taps to the same setting and the three fine taps to the same setting (see Section 6.3.).

6.1.14 Alternately, the approximate tap setting could be calculated by using the following procedure.

6.1.14.1 Energize the rectifier and take the following measurements:

- DC volts
- DC amperes
- AC secondary volts
- Tap setting

6.1.14.2 Calculate the DC load resistance (anode to structure) using Equation (6.1):

$$R_a = \frac{(E_{DC} - \text{Back EMF})}{I_{DC}}, \tag{6.1}$$

where

R_a	anode to structure resistance (ohms)
E_{DC}	DC voltage output (volts)
Back EMF	open-circuit DC voltage between anode and structure (before energizing; volts)
I_{DC}	DC current output (amperes)

Example: If the rectifier output is 10.0 V_{DC} and 4.0 A_{DC} and the open-circuit back EMF is 2.0 V_{DC}, then

$$R_a = \frac{(10.0\ V_{DC} - 2.0\ V_{DC})}{4.0\ A_{DC}}$$

$$R_a = 2.0\,\Omega.$$

Note that it is incorrect to simply divide the output voltage by the output current without accounting for the back EMF as, in this example, it would give a wrong answer of 2.5 Ω for resistance.

6.1.14.3 Knowing the required current, the DC output voltage can then be calculated using Equation (6.2):

$$E_{\mathrm{DC}} = I_{\mathrm{DC}} R_a + \text{Back EMF}, \tag{6.2}$$

where

E_{DC}	DC voltage output (volts)
I_{DC}	DC current required (amperes)
R_a	resistance between anode and structure (ohms)
Back EMF	open-circuit DC voltage between anode and structure (volts)

6.1.14.4 Knowing the secondary AC voltage between the taps, as measured in 6.1.9, and allowing an additional voltage drop through the rectifying elements, the AC tap setting can be predicted.

For example, the AC voltage between fine taps is 2.0 $\mathrm{V_{AC}}$, and the AC voltage between the coarse taps is 10.0 $\mathrm{V_{AC}}$. Assume, to start, that there is a 15% voltage drop through the rectifying elements. In this case, if the DC voltage required is 12.0 $\mathrm{V_{DC}}$, the AC voltage must then be

$$E_{\mathrm{AC}} = 12.0\,\mathrm{V_{DC}} \times 1.15\,(15\%) = 13.8\,\mathrm{V_{AC}}. \tag{6.3}$$

Without back EMF, the closest setting would be one coarse tap at 10 $\mathrm{V_{AC}}$ each, and two fine taps at 2 $\mathrm{V_{AC}}$ are required, or the trial setting would be coarse 2 (B), fine 2.

6.2 Adjusting an Operating, Single-Phase, Tap-Type Rectifier

6.2.1 Determine the reason that the current has to be changed and confirm that it is safe to do so. For example, it would not be appropriate to increase the rectifier to protect a portion of the system isolated by a broken bond.

6.2.2 Turn the rectifier off before adjusting the taps for each adjustment.

6.2.3 Adjust the voltage, and, therefore, the current, by changing the fine and coarse taps until the required current has been reached (see Section 6.1.12).

6.2.4 If an unknown reduction of current is needed, such as when a fuse or circuit breaker is blown, then follow the procedure in Section 7.0.

6.3 Adjusting a Three-Phase, Tap-Type Rectifier

6.3.1 The procedure described in Section 6.1 or 6.2 is to be completed, with the following exceptions.

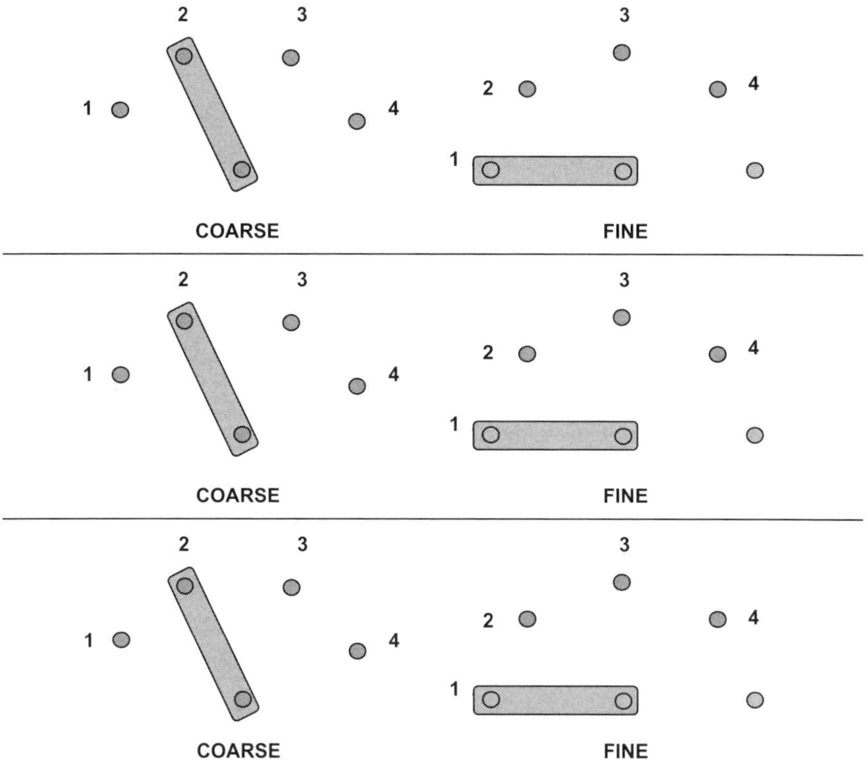

Figure 6.3 A sample setting for a three-phase rectifier is shown, showing the same coarse taps and the same fine taps (recorded as 3× C-2 F-1).

6.3.2 The primary AC supply voltage between each of the three phases is to be measured. These values are expected to be similar if the phases are balanced.

6.3.3 There will be three duplicate coarse and fine settings: one group for each phase (Figure 6.3). Measure the AC voltage within the taps for each group only.

6.3.4 When adjusting the taps, the coarse taps are to be set at the same number or letter, and similarly, the fine taps are to be set at the same number or letter (Figure 6.3). Otherwise, the phases will be out of balance.

6.4 Adjusting a Constant Current Rectifier

6.4.1 A constant current rectifier maintains the set current output by varying the DC voltage within the rating of the rectifier to compensate for a change in the external resistance (anode), as shown in Figure 6.4.

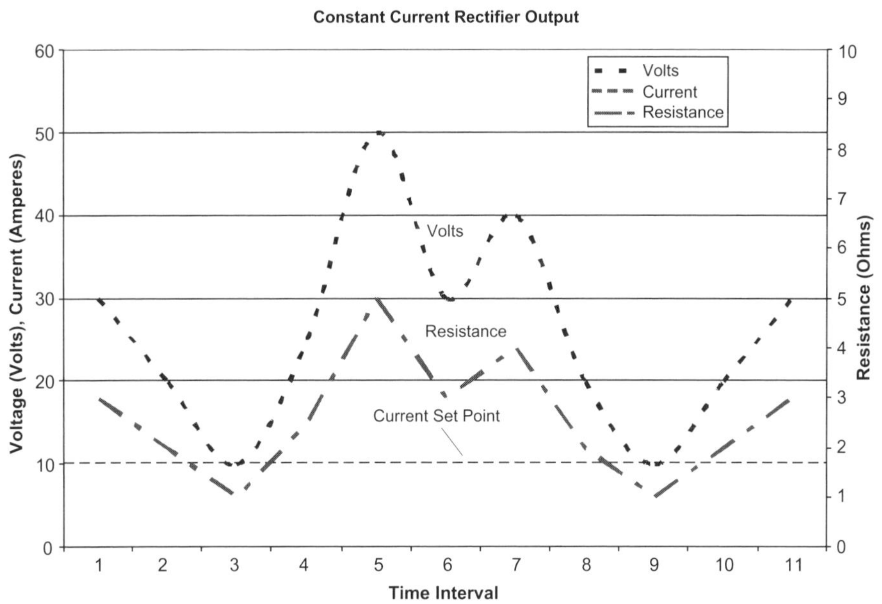

Figure 6.4 Rectifier voltage automatically adjusts with change in resistance to maintain a constant current.

6.4.2 A constant current rectifier may be controlled by either a saturable reactor or silicon-controlled diodes.

6.4.3 Follow the manufacturer's instructions to adjust the current.

6.4.4 Generally, a control feature should be labeled that will increase the rectifier current when turned in a clockwise direction and decrease the current when rotated in an anticlockwise direction. The current will remain constant at the preset value up to the rating of the rectifier.

6.4.5 If the rectifier also has a voltage adjustment, set this control to the maximum value, unless a specific upper limit has been predetermined.

6.4.6 Note that demanding the same current as the anode bed resistance continues to increase may cause accelerated drying and premature failure of the anodes.

6.5 Adjusting a Constant Potential Rectifier

6.5.1 The constant potential rectifier will maintain a stable potential between the structure and a stationary reference electrode by adjusting the current appropriately within the rating of the rectifier, as shown by Figure 6.5. When the potential starts to go more electropositive than the set point, the current will be increased by a corresponding voltage increase, and if the potential goes more electronegative, then the rectifier decreases the current, and, thus, the potential stays constant. The resistance is shown as constant in Figure 6.5, but within the limits of the rectifier, a change in resistance would also be compensated.

6.5.2 Make sure that the stationary reference electrode used matches the control setting (i.e., copper-copper sulfate (CSE) or zinc).

6.5.3 Normally, rotate the potential control adjustment in a clockwise direction to make the structure-to-electrolyte potential more electronegative and in a counterclockwise direction for a less electronegative potential.

6.5.4 If the rectifier has a voltage control adjustment, set this at the maximum value, unless a specific upper voltage limit has been predetermined.

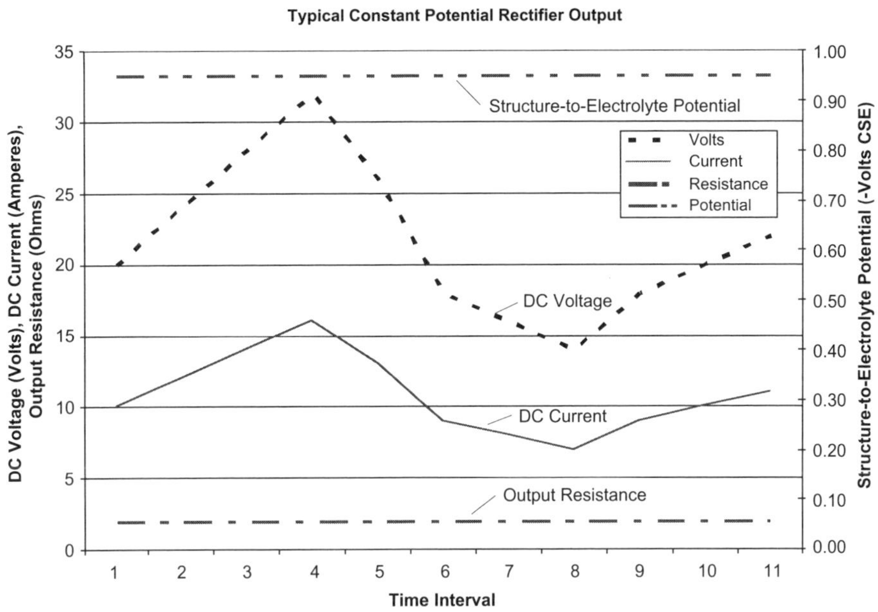

Figure 6.5 Constant potential is maintained by rectifier changing voltage output to change current.

6.5.5 If the rectifier has a current control adjustment, set this at the maximum value, unless a specific upper current limit has been predetermined.

7.0 INSPECTION PROCEDURE

7.1 General

7.1.1 Review precautions (Section 4.0). Do not complete tests beyond your qualifications.

7.1.2 Test the rectifier case for a hazardous voltage before touching.

7.1.3 Failures can occur in any of the rectifier components, connecting wires, or connections themselves or in the external circuit.

7.2 Preliminary Inspection

7.2.1 Before touching the rectifier, measure the case voltage-to-ground and de-energize it from the AC supply disconnect, if hazardous.

Alternately, test the case with a voltage alert type of instrument that indicates a hazardous voltage with a light.

7.2.2 Inspect the rectifier mounting and case for damage.

7.2.3 Confirm that the rectifier case is properly grounded.

7.2.4 Listen for unusual sounds such as arcing or high vibrations.

7.2.5 Note the operating temperature of the rectifier.

7.2.5.1 If it is an air-cooled rectifier, ensure that the screens are clear and that the rectifier is in a location to allow good air circulation. Clean the screens where clogged and/or replace a screen if it has holes that might allow insects, reptiles, or rodents to gain access. Relocate the rectifier if it is in an area that will not allow good air circulation.

7.2.5.2 If it is an oil-cooled rectifier, ensure that the sides are clear to allow heat to radiate from the sides and top. Remove any article that has been wedged against the case and any heat-insulating coating to reduce heat radiation. If the oil is not clear, replace it. Check for water accumulation at the bottom and drain periodically.

7.2.6 Look for signs of high temperature such as discoloration of copper terminals or arc burns on the panel. Do not touch while energized or still hot.

7.2.7 With the rectifier de-energized, clean and tighten all connections.

7.2.8 Record the nameplate data.

7.2.9 Record the present DC output settings.

7.2.10 If it is a constant voltage rectifier, record the secondary tap AC voltage and compare with the DC output voltage. (The values for a single-phase rectifier should be within ~15%.)

7.2.11 If the DC voltage is less than one-half of the secondary AC voltage, then one-half waving is suspected (see Section 8.0).

7.2.12 Check for poor connections between the secondary taps and the DC output.

7.2.13 Hot fuse terminals will cause the fuse to blow prematurely. Clean and tighten if it is a bolted type. If it is a spring-type holder, replace the holder as the spring tension will have been lost due to overheating.

7.3 Inspection

7.3.1 Routine Monitoring

7.3.1.1 The more frequent the monitoring of the rectifier, the less time the structure will be without protection when a failure occurs; therefore, monthly readings are recommended at a minimum, unless another means of confirming protection is in place. For example, if a rectifier goes off just after a bimonthly inspection once each year, then the structure has been without protection for a total of 1 year every 6 years.

7.3.1.2 As a minimum, record the rectifier DC current and DC voltage output and the tap setting or control setting. The latter may explain a change in output. A continuing plot of circuit resistance will indicate when the anodes are starting to fail.

7.3.1.3 If a power meter is available, record the kilowatt-hours (KWH) reading, the number of revolutions per minute, and the Kh factor on the face of the meter.

7.3.2 Annual Inspection

A detailed inspection of the rectifier should be completed at least annually, with the following information collected:

7.3.2.1 Record the nameplate data or confirm that previous information is still accurate.

7.3.2.2 Determine the DC voltage and DC current output using a portable meter and calibrate the panel meters, if applicable.

7.3.2.3 Determine the transformer secondary (stepped down) coarse-to-fine tap volts$_{AC}$.

7.3.2.4 Compare the DC volts to the tap volts$_{AC}$. If the two voltages are not within 15% to 20%, then inspect the rectifier bridge circuit and, if necessary, arrange for repair.

7.3.2.5 Find the DC volts between the positive and negative terminals with the DC power source temporarily off. This voltage, referred to as back EMF, will be used in calculations to determine anode bed resistance.

7.3.2.6 Calibrate the voltmeter and ammeter either by adjusting to the correct value or recording the discrepancy.

7.3.2.7 Determine the voltmeter's accuracy by comparing the meter's reading to a portable voltmeter.[2]

7.3.2.8 The panel ammeter's accuracy can be determined by comparing the ammeter's reading to that calculated from the millivolt reading obtained across the calibrated shunt. If using the calibrated shunt, measure the millivolt drop across the shunt terminals with a portable millivoltmeter. Convert the millivolt reading to amperes.[3] One method of doing this is to use the following ratio:

$$I_{\text{calculated}} = I_1(V_{\text{measured}}/V_1), \tag{7.1}$$

where

$I_{\text{calculated}}$	calculated ampere output (amperes)
V_1	shunt's voltage rating (millivolts)
I_1	shunt's ampere rating (amperes)
V_{measured}	measured voltage drop across shunt (millivolts)

7.3.3 With the power off, carefully inspect for thermally hot terminals. In addition to looking for burn marks, melted mounting boards, and so on, use the back-of–the-hand method to sense heat. This involves placing the back of your hand in close proximity to the components in an attempt to sense whether heat is emanating from the component. Care must be taken to ensure that this is done without actually touching the component to avoid burns.

7.3.4 If it is an oil-cooled unit, the oil level and clarity of the oil of oil-immersed units is to be recorded. If the oil level is low, then it must be filled. If the components of the rectifier cannot be easily seen through the oil because it has become murky, the oil needs to be replaced. Replacement oil should be in accordance with manufacturer recommendations.

---

[2] References to portable meters imply that the portable meters have recently been calibrated and are accurate.

[3] *See* Cathodic Protection Procedure No. 3: Direct Current Measurements.

7.3.5 The vents of air-cooled units must be clear and screens in place. Nests and debris must be removed from the cabinets. Exercise caution as the nests may house insects, snakes, or other creatures.

7.3.6 Seasonal changes in anode bed resistance (see Equation [7.2] to calculate anode bed resistance) may occur throughout a given year, and they are often repeated in subsequent years.

$$R_a = \frac{(E_{\mathrm{DC}} - \text{Back EMF})}{I_{\mathrm{DC}}}, \tag{7.2}$$

where

R_a	anode to structure resistance (ohms)
E_{DC}	DC voltage output (volts)
Back EMF	open-circuit DC voltage between anode and structure (galvanic difference between steel and carbon in coke breeze backfill measured with rectifier off and in this case normally between 1.0 and 2.0 V_{DC}) (volts)
I_{DC}	DC current output (amperes)

A continuous or very rapid increase in resistance over time indicates anode deterioration. A close interval potential survey can be completed over shallow anodes. An increase in the voltage gradient will indicate an operating anode (Section 8.5).

7.3.7 Where applicable, the rectifier efficiency can be calculated using Equations (7.3) and (7.4). Note that Equation 7.3 is an approximation as the power factor is not considered in the AC watts ($V_{\mathrm{AC}} I_{\mathrm{AC}}$).

$$\text{Eff} = \frac{V_{\mathrm{DC}} I_{\mathrm{DC}}}{V_{\mathrm{AC}} I_{\mathrm{AC}}} \times 100, \tag{7.3}$$

where

Eff	efficiency (%)
V_{DC}	DC voltage output (volts)
I_{DC}	DC current output (amperes)
V_{AC}	AC voltage output (volts)
I_{AC}	AC current output (amperes)

If a power meter is in place, calculate the rectifier efficiency using Equation (7.4). The K_h factor in this equation is taken from the power meter nameplate

$$\mathrm{Eff} = \frac{V_{\mathrm{DC}} I_{\mathrm{DC}} T}{3600\, K_h n} \times 100, \tag{7.4}$$

where

Eff	efficiency (%)
V_{DC}	DC voltage output (volts)
I_{DC}	DC current output (amperes)
K_h	power meter factor
n	number of revolutions
T	time of the revolutions

7.3.8 The rectifier efficiency will vary, depending on the percentage of rated voltage and current output at which it is operating.[9.2] Maximum efficiency is achieved at rated voltage and current output.

7.3.9 The average rectifier efficiency over time can be calculated by Equation (7.5):

$$\mathrm{Eff}_{\mathrm{ave}} = \frac{V_{\mathrm{DC}} I_{\mathrm{DC}} T_{\mathrm{DC}}}{(\mathrm{KWH}_2 - \mathrm{KWH}_1)1{,}000} \times 100, \tag{7.5}$$

where

$\mathrm{Eff}_{\mathrm{ave}}$	average efficiency for time (T_{DC})(%)
V_{DC}	DC voltage output (volts)
I_{DC}	DC current output (amperes)
KWH_1	power consumption read at beginning of time interval (KWH)
KWH_2	power consumption read at end of time interval (KWH)
T_{DC}	assumed time for DC between start and end of KWH readings (hours)

A higher efficiency from Equation (7.5) than that calculated from Equations (7.3) or (7.4) indicates a rectifier outage as the DC power consumption is assumed to be continuous and the AC consumption is the actual amount. The length of time that the rectifier has been off can be calculated from Equation (7.5) using the efficiency when the rectifier is operating normally and solving for time (T_{DC}; see Equation [7.6], rearranged from Equation [7.5]):

$$T_{DC\ actual} = \frac{Eff_{inst}(KWH_2 - KWH_1) \times 1{,}000}{V_{DC} I_{DC} 100}, \tag{7.6}$$

where

Eff_{inst}	instant efficiency or the average efficiency over time (%)
V_{DC}	DC voltage output (volts)
I_{DC}	DC current output (amperes)
KWH_1	power consumption read at beginning of time interval (KWH)
KWH_2	power consumption read at end of time interval (KWH)
$T_{DC\ actual}$	actual operating time for DC between start and end of KWH readings (hours)

7.3.10 During a rectifier inspection, a structure-to-electrolyte potential must be taken near the rectifier both to confirm proper polarity of the rectifier connections by an electronegative change in potentials when the rectifier comes on and to note the level of protection on the structure. This should also be completed as part of the monthly rectifier monitoring and annual rectifier inspections or any rectifier maintenance and relocation.

8.0 TROUBLESHOOTING PROCEDURE

8.1 General

8.1.1 The source of trouble may be either in the AC supply or in the external DC circuit, in addition to the rectifier.[9.3,9.4] Several different approaches can be taken to locate the source of trouble. This approach to troubleshooting a rectifier (Figure 8.1) attempts

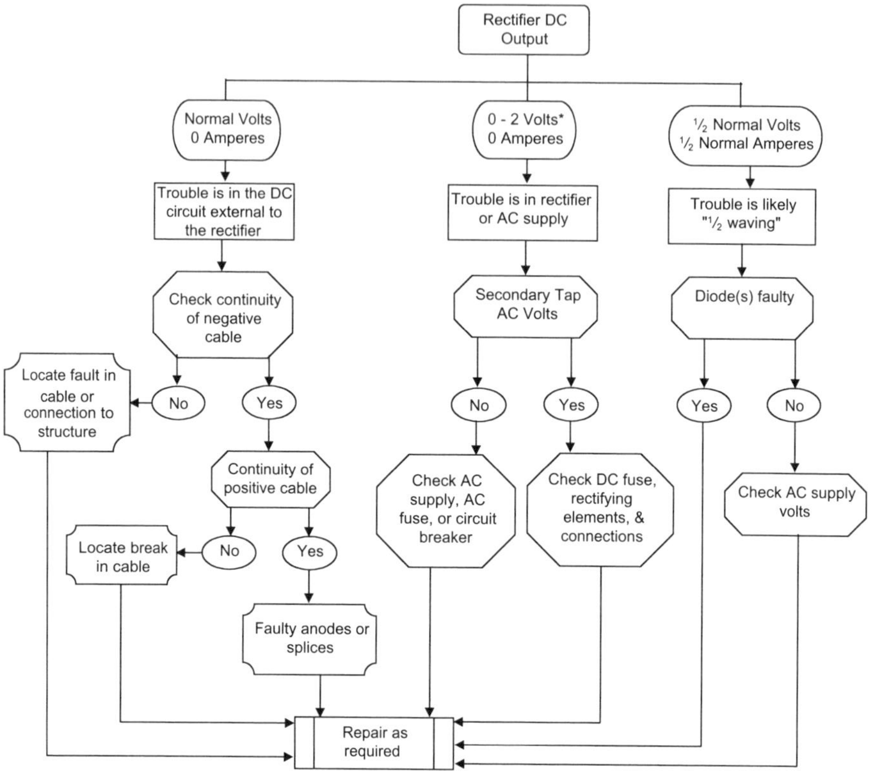

* If 2 volts or less, confirm that it is a back EMF by disconnecting a DC cable.

Figure 8.1 One approach to locate trouble in a cathodic protection circuit is shown.

to determine quickly the general area of trouble and then focus on that area with more detailed troubleshooting (see also Table 8.1). If the trouble is not immediately obvious, then it is best to systematically isolate the components until the defective part is isolated.

8.1.2 If there is normal DC voltage and 0 A, the trouble is in the external circuit and may be due to

- Deteriorated anodes
- Faulty cable splices
- Broken cable
- Faulty structure connections

Table 8.1 Summary of DC Power Supply Troubleshooting

DC Power Source								
Volts	Amperes	AC Supply	Secondary AC Voltage	Breaker Trips/Fuse Blown	Structure-to-Electrolyte Potentials	Trouble Suspected	Tests/Inspection	Remedy
0 to 2*	0	0	0	–	P+	No AC supply	Test AC supply, inspection power supply, AC breakers, or power line fuse.	Arrange to have AC supply restored.
0 to 2*	0	V	0	Yes	P+	Taps set too high, faulty breaker, or short in DC power source, including surge protectors or external circuit.	Disconnect DC output cable. Measure external resistance and determine if taps were set too high. If OK, then, with power off, disconnect breaker output and surge protection. Turn on AC and breaker and measure AC volts out of breaker. If OK turn off AC, reconnect breaker, and turn on. If breaker trips then there is a short in the transformer or the circuit breaker is weak. If OK then test the rectifying elements and the connecting wires for a short.	Lower tap setting or repair or replace shorted component.

(*continued*)

Table 8.1 *(continued)*

DC Power Source								
Volts	Amperes	AC Supply	Secondary AC Voltage	Breaker Trips/Fuse Blown	Structure-to-Electrolyte Potentials	Trouble Suspected	Tests/Inspection	Remedy
0 to 2*	0	V	0	No	P+	Open circuit in breaker or transformer or secondary taps.	Test AC volts in and out of circuit breaker. If good then open circuit exists in transformer or secondary taps.	Replace breaker, transformer, or repair secondary taps as required.
0 to 2*	0	V	V	Breaker OK	P+	AC fuse, failed rectifying element(s), DC fuse, or faulty connections in power supply.	Test AC and DC fuses if applicable. If fuse is blown, then disconnect output cable and test external resistance. Lower taps if necessary. Otherwise, test power supply wire and connections and rectifying elements.	Replace or repair as required.
$\sim\frac{1}{2}$ V	$\sim\frac{1}{2}$ A	V	V	No	P+	$\frac{1}{2}$ wave output, i.e., one leg of the rectifier bridge is open or badly deteriorated if selenium	Compare secondary AC volts with DC volts. If ratio of DC/AC volts is $\sim\frac{1}{2}$, then rectifying element is faulty.	Replace rectifying element as necessary.

0	A	V	V	No	P	Faulty voltmeter	Measure output volts	Replace or repair as required.
V	0	V	V	No	P+	Fault in external circuit	Test cable continuity, structure connections and anodes.	Replace or repair as required.
V	A+	V	V	No	P−	Anode resistance reduced	Measure external resistance and reduce secondary taps.	Adjust taps.
V±	A±	V	V	No	P±	Poor connections / meters	Test connections in power supply and external circuit.	Repair as required.

Legend:

V Normal volts
A Normal amperes
V+ Greater-than-normal volts
V− Lower-than-normal volts
A+ Greater-than-normal amperes
A− Lower-than-normal amperes
P+ Structure-to-electrolyte potential more electropositive than normal
P− Structure-to-electrolyte potential more electronegative than normal
V± A± P± Volts, amperes, and structure-to-electrolyte potentials varying
* ~2 volts may be due to galvanic difference between the steel and anodes or the carbon in the coke breeze and not an indicator of power.

8.1.3 If the rectifier output reads 0 to ±2 V_{DC} and 0 A_{DC}, the trouble is in the rectifier or the AC supply. The ±2 V is likely the back EMF due to the galvanic difference between the metal in the structure and the carbon in the coke breeze. Disconnect the cable and measure the DC voltage between the cables. The DC voltage in the rectifier will go to 0 V_{DC}, while the DC voltage between the positive and negative cable will remain the same if it is due to the back EMF.

8.2 External (Structure-Anode Circuit)

If the negative cable is good and the positive cable is continuous to the anodes, then either the anode leads, the anode connection, the anode has failed or the anodes have deteriorated. The latter usually occurs over time and can be seen as a gradual increase in resistance to start and then a rapid increase as the last remaining anodes fail.

8.2.1 Look for signs of excavation crossing the cables or the anodes.

8.2.2 Determine if the open circuit is in the positive or negative circuit by one of the following procedures:

8.2.2.1 Install a ground rod or an isolated ground, connect to the positive, and record any current response in the ammeter (Figure 8.2, connection "C"). If there is a break in the positive circuit, there will be a small response in the rectifier ammeter, but there will be no response if there is a break in the negative cable.

8.2.2.2 Disconnect the ground from the positive, or use a separate ground, and connect the ground to the negative (Figure 8.2, connection "B") and record any response in the rectifier ammeter. If there is a break in the positive cable, there will be no response, but there will be a small response if the break is in the negative cable or connection. Note that the current response is not expected to be the same as when operating normally.

8.2.2.3 Connect a temporary cable from the rectifier negative to a test station or directly to the structure (Figure 8.2, connection "A"). If the break is in the negative cable or

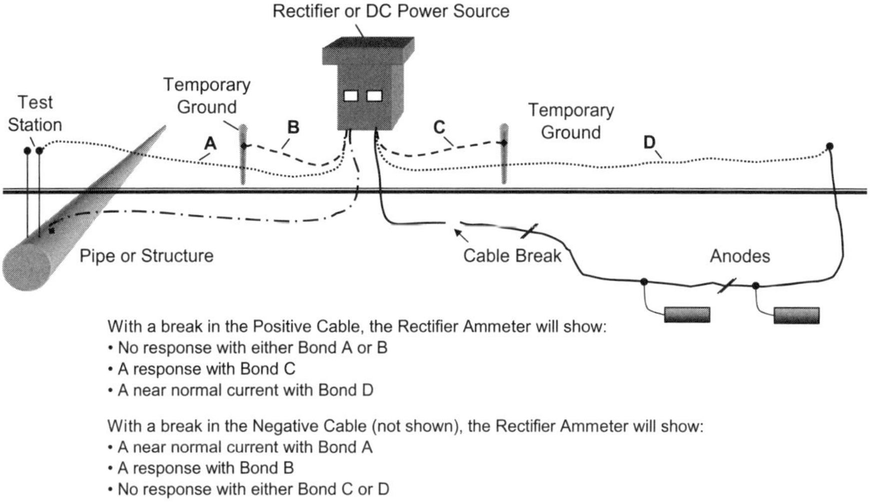

Figure 8.2 Testing the negative and positive cables by noting a rectifier current response is shown.

connection, the rectifier ammeter will indicate a near-normal output, depending on the resistance of the temporary wire, but there will be no response if the break is in the positive cable or anodes.

8.2.2.4 Connect a temporary wire to the end of the anode bed or anode junction box, if available (Figure 8.2, connection "D"). If the break is in the negative cable or structure connection, the rectifier ammeter will indicate no response, but there will be a near-normal output, depending on the resistance of the temporary wire, if the break is in the positive cable or anodes.

8.2.3 Alternately, disconnect the negative cable and measure the resistance between the cable end and the structure (Figure 8.3). Note that when using an AC resistance meter, do not leave any wire on the reel because this will create varying impedance in the test circuit due to the inductance in the wire coil.

8.2.4 Measure the resistance of the wires used in the test and subtract this value from the first resistance measurement. The answer will be the resistance of the negative cable and structure connection.

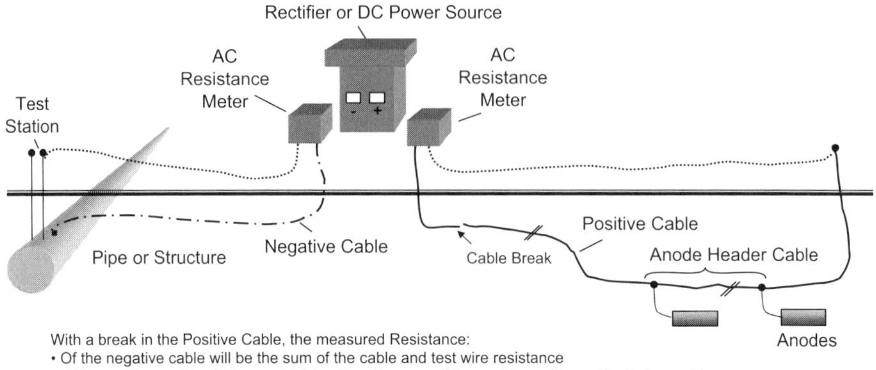

With a break in the Positive Cable, the measured Resistance:
- Of the negative cable will be the sum of the cable and test wire resistance
- Of the positive cable will be much higher than the sum of the positive cable and test wire resistances
- May be a combination of different anode resistances if the break is in the anode header cable (Resistance is measured from one group of anodes to another through the earth)

With a break in the Negative Cable or connection, the Resistance:
- Of the negative cable will be very high
- Of the positive cable will be the sum of the positive cable and test wire resistances

Figure 8.3 Measuring resistance of negative and positive cables is shown.

8.2.5 If a connection can be made to the positive cable at the anode bed, either at the end or at a junction box, disconnect the positive cable and measure the cable resistance (Figure 8.3). Again, subtract the resistance of the test wires. Do not use a DC ohmmeter in this case. If the break is between the anodes, there will be a voltage difference across the break that will either add to or oppose the meter voltage, causing an error in the reading.

8.2.5.1 If the break is in the positive cable, the resistance of the negative cable measurement will be the sum of the negative cable resistance and the test wire resistance. The resistance of the positive cable will be very high if it is before the anode bed, but if it is in the anode header cable, it will be a combination of the resistance to earth of the groups of anodes on each side of the header cable break.

8.2.5.2 If the break is in the negative cable, the resistance of the negative cable measurement will be very high, but the measured resistance of the positive cable will be the sum of the positive cable and the test wire resistances.

8.3 Cable Tracing

Using a conductive pipe and cable locator, connect the transmitter between a ground and either the negative or positive cable and trace the signal. Note that the signal will normally extend a short distance beyond the break at a lower strength. If there are one or more cables in parallel in the ditch and the other(s) is continuous, the signal will be induced on the second cable and it can continue past the break on the good cable. Another method to locate the break is then required.

8.4 One-Half Method

8.4.1 Excavate the cable at approximately the halfway point and expose a small portion of the copper cable. Measure the continuity or resistance from that point to the rectifier. If continuous, the break is beyond that point, and if not continuous, the break is in between the points connected in the test.

8.4.2 Excavate the section of cable that is not electrically continuous at the halfway point between the former test location and the end. Measure the continuity or resistance between this point and the former halfway point and also to the end.

8.4.3 Repeat this process until the location of the break is determined within a small area. Often, a pull on the cable will yield at the break when it is nearby.

8.4.4 Repair the break and cover all copper exposed during the test with a water barrier material.

8.5 Failing Anodes

8.5.1 Typically, failing anodes can be detected by observing a gradual increase in resistance over time. This resistance increase will start gradually but will accelerate as fewer anodes are left operating.

8.5.2 If the fault appears to be in a shallow anode bed that still has some current output, conduct a close interval potential survey over the anode area. Operating anodes will be depicted by the increase in the voltage gradient (*IR* drop) centered through each anode. There will be minimal change in potential through the depleted anode section(s).

8.6 Transformer-Rectifier

8.6.1 Alternating Current Supply and Primary Transformer Taps

8.6.1.1 Measure the AC voltage and compare with the rectifier rating. If it is a dual–AC input rectifier, confirm that the taps are set correctly for the AC supply.

8.6.1.2 If, for example, the rectifier is a 115/230-V_{AC} unit and is supplied with 115 V_{AC}, but the AC input link bars are connected for a 230-V_{AC} supply, the maximum DC voltage output will be one-half that of rated. This may have been done intentionally in some cases to gain finer control of DC output with low resistance anode beds, but a label should be placed on the rectifier and AC input in this case as this may appear to be a code violation. If the rectifier is equipped with electronics, it must be connected to the rated AC supply.

8.6.1.3 If, for example, the rectifier input taps are set for 115 V_{AC} but the unit is supplied with 230 V_{AC}, damage to the transformer will likely occur.

8.6.2 Alternating Current Surge Protection

8.6.2.1 Check the condition of the AC surge protection. Signs of excessive heat can usually be seen if it has failed. It is intended to protect the rectifier from either fault current or lightning surges coming from the AC supply.

8.6.2.2 As many will fail in a shorted condition, the breaker will be off, and an attempt to reset it will cause an immediate release.

8.6.2.3 Disconnect the AC surge protection and reset the AC circuit breaker. If the breaker stays on, then replace the AC surge protection. If it releases, then there is a short in the rectifier or in the external circuit.

8.6.2.4 Note that the rectifier is, of course, not protected from surges if the surge protection is left off.

8.6.3 Alternating Current Circuit Breaker

8.6.3.1 The breaker may fail in a shorted position where the contacts have welded together due to arcing, or alternatively, the trip element may fail to hold the contacts

Table 8.2 Possible AC Circuit Breaker Conditions

Breaker Off		Breaker On		
AC In	AC Out	AC In	AC Out	Problem
Normal V_{AC}	0 V_{AC}	Normal V_{AC}	Normal V_{AC}	Good condition
Normal V_{AC}	Normal V_{AC}	Normal V_{AC}	Normal V_{AC}	Contacts welded closed
Normal V_{AC}	0 V_{AC}	Normal V_{AC}	0 V_{AC} (trips)	Faulty breaker or short in breaker, surge protection, rectifier, or external circuit

closed. Confirm that the AC power is indeed disconnected by taking an AC voltage measurement.

8.6.3.2 Measure the AC voltage "in" to the circuit breaker (normally at the top) and "out" with both the circuit breaker on and off. The conditions that may exist are given in Table 8.2.

8.6.4 Transformer

8.6.4.1 Groups of primary windings can be placed in series or in parallel in the case of a dual-AC voltage supply. This is achieved by taps near the AC supply terminals. Confirm that the taps are set for the AC supply voltage (see Section 8.6.1).

8.6.5 Secondary Windings

8.6.5.1 Measure the AC voltage between taps. The sum of the fine tap voltages should approximate the AC voltage between each of the coarse taps. Confirm that all taps are properly connected by measuring an ever increasing voltage as the taps increase with one meter lead left on fine 1.

8.6.5.2 A tap-to-tap measurement of 0 V_{AC} indicates a faulty secondary tap, or if there is not a continuous increase in the AC voltage as the taps increase, the taps are likely mislabeled.

8.7 Alternating Current Fuse

8.7.1 If a fuse is installed between the transformer secondary taps and the rectifying element and an AC voltage exists at the secondary taps but not at the AC input to the rectifying elements, remove and test the AC fuse or check the connections from the taps through to the rectifying element.

8.7.2 A blown fuse indicates either an overload in the rectifying elements, external circuit, or hot contacts in the fuse holder.

8.7.3 Inspect the fuse holder contacts for signs of overheating. Replace a spring-type fuse holder with signs of overheating. Clean a bolted-type fuse holder and ensure that there is a tight connection.

8.7.4 Before replacing the fuse, determine the cause of the failure using one of the following methods.

8.7.4.1 Test the output circuit resistance and calculate the current output at the secondary tap setting. Adjust taps (see Section 6.1.12) if the DC current is too high.

8.7.4.2 Test the rectifying elements for a shorted diode and replace, as required.

8.7.4.3 Inspect for signs of arcing to the rectifier case. Determine cause and repair, as necessary.

8.8 Rectifying Element

8.8.1 If the rectifying element consists of selenium, the failure will generally be in the form of reduced efficiency, with signs of arc burns on the plates. A catastrophic failure will result in burned and warped plates that are loose in the stack.

8.8.2 A silicon diode will either operate or fail in either an open circuit or shorted condition.

8.8.3 A sign of partial failure of one leg of the bridge circuit, or one-half waving, is when the DC voltage output is less than one-half that of the AC secondary tap voltage.

8.8.4 Either remove the rectifying elements and test each part independently, or perform the test described in Section 8.8.5. If it is a diode bridge, use a meter with a diode test circuit to test the forward and reverse bias. Typical readings from a diode tester are given in Table 8.3.

8.8.5 The diode bridge circuit can be tested in place using the following procedure and as illustrated in Figure 8.4. Typical readings from the diode tester are given in Table 8.3.

Table 8.3 Diode Meter Readings Indicating Diode Condition

Diode Condition	Forward Reading (V)	Reverse Reading (V)
Good	~0.5	OL
Shorted	~0	~0
Open circuit	OL	OL

8.8.5.1 Turn off and lock out/tag out the rectifier.
8.8.5.2 Confirm that all rectifier fuses are in good condition, if applicable.
8.8.5.3 Remove at least one tap bar.
8.8.5.4 Remove at least the positive or negative cable. (If more than one positive or negative cable exists, then all cables connected to the terminal must be removed.)
8.8.5.5 Set the meter to diode test.
8.8.5.6 With the positive meter lead on the tap and the negative lead on the rectifier positive terminal, the reading should be approximately 0.5 V.
8.8.5.7 Reverse the meter leads and connect to the same terminals, and the reading should be "OL."

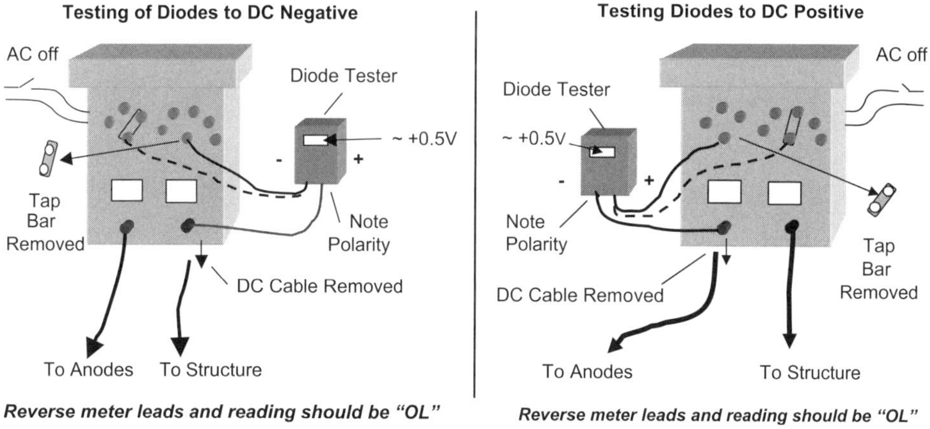

Figure 8.4 A technique for testing rectifying elements in place is shown.

Note: Power must be off and locked out. If open circuit is indicated, check for blown fuses or bad connections.

8.8.5.8 With the positive meter lead on the rectifier negative terminal and the negative lead on the AC tap center terminal, the reading should be approximately 0.5 V.

8.8.5.9 Reverse the meter leads and connect to the same terminals, and the reading should be "OL."

8.9 Direct Current Fuse

8.9.1 If a fuse is installed between the rectifying element and the DC output, and a DC voltage exists at the DC output of the rectifying elements but not at the DC output, remove and test the DC fuse. Another test is to measure the DC voltage across the fuse. There should be essentially no voltage if the fuse is good.

8.9.2 A blown fuse indicates an overload in either the rectifier circuit, the external circuit, or the hot contacts in the fuse holder.

8.9.3 Inspect the fuse holder contacts for signs of overheating. Replace a spring-type fuse holder with signs of overheating. Clean a bolted-type fuse holder and ensure that there is a tight connection.

8.9.4 Before replacing the fuse, determine the cause of the failure using the following steps.

8.9.4.1 Test the output circuit resistance and calculate the current output at the secondary tap setting. (See Section 6.1.14.) Adjust taps if the DC current is too high.

8.9.4.2 Test the rectifying elements for a shorted diode and replace, as required.

8.9.4.3 Inspect for signs of arcing to the rectifier case. Determine cause and repair, as necessary.

8.10 Meters

8.10.1 The DC output of most rectifiers is monitored by a DC voltmeter and a DC ammeter with either an internal or external shunt.

8.10.2 Faulty meters can give a false indication of trouble. For example, a failed ammeter with accompanying DC voltage suggests that there is trouble in the external circuit, while the trouble is in the ammeter itself.

8.10.3 Confirm the DC output measurements with a portable meter before investigating further.

8.11 Direct Current Surge Protection

8.11.1 Inspect for signs of arcing and disconnect before re-energizing the rectifier.

8.11.2 A summary of the troubleshooting of a common transformer-rectifier circuit breaker is given in Table 8.2.

8.12 Final Testing

8.12.1 Measure the polarity of the DC output and confirm that the positive is connected to the anodes before leaving any cathodic protection power source on after maintenance; confirmation of the correct polarity must be made.

8.12.2 Compare the OFF structure-to-electrolyte potential to the ON structure-to-electrolyte potential. The values should become more electronegative after the rectifier is turned on. If not, do not leave it energized, and investigate the rectifier DC output polarity connections. Reversed rectifier connections will cause accelerated corrosion damage to the structure.

8.12.3 Ensure that a drawing showing the location of underground wiring, polarity, and anodes is provided inside the rectifier cabinet or in a location nearby. This is a regulatory requirement in some countries.[9.5]

9.0 REFERENCES

9.1 Appalachian Underground Corrosion Short Course, 2000 Revision, *Intermediate Course*, Chapter 7.

9.2 A.W. Peabody, *Control of Pipeline Corrosion*, 2nd ed., ed. R.L. Bianchetti (Houston, TX: NACE, 2001), p. 164.

9.3 W. von Baeckmann, W. Schwenk, W. Prinz, *Handbook of Corrosion Protection*, 3rd ed. (Houston, Texas: Gulf Publishing Company, 1997), p. 229.

9.4 M.E. Parker, E.G. Peattie, *Pipe Line Corrosion and Cathodic Protection*, 3rd ed. (Houston, London, Paris, Tokyo: Gulf Publishing Company, Book Division, June 1995), p. 126.

9.5 CSA Standard C22.1-02, *Canadian Electrical Code, Part 1* (Ontario, Canada: Canadian Standards Association, Etobicoke, 2002), p. 226.

CATHODIC PROTECTION PROCEDURE NO. 2

Structure-to-Electrolyte Potential Measurement 2

1.0 INTRODUCTION

The purpose of this procedure is to measure a potential difference between a buried or immersed metallic structure and a reference electrode in contact with that same electrolyte.

Structure-to-electrolyte potentials are taken for many different purposes, such as tests for the following:

- Confirming cathodic protection (CP) criteria
- Electrical isolation
- Continuity
- Road and railroad casing
- Shielding
- Stray current
- Coating evaluation

Each of these tests is covered in detail by another CP procedure module. The number and location of structure-to-electrolyte potential measurements will be determined by the objective of the particular survey.

2.0 TOOLS AND EQUIPMENT

- Voltmeter:
  - High input impedance (10 MΩ minimum) direct current (DC) voltmeter to measure from 0 to 4,000 mV and up to 20 V

    - Optional DC voltmeter with a variable-input impedance from 10 to 200 MΩ or an add-on to change the internal voltmeter impedance
    - Optional DC voltmeter/data logger with a memory programmed to store value, units, polarity, time, and date, either on command or at time intervals
    - Note that voltmeters of less than 10 MΩ require a potentiometer circuit
- Reference electrode:
    - Copper-copper sulfate
    - Silver-silver chloride
    - Optional reference electrode extension
- Test leads complete with electrically insulated spring clips or connectors
- Wire reel
- Alternating current (AC) voltmeter

3.0 SAFETY EQUIPMENT

- Standard safety equipment as required by the company's safety manual and regulations
- Electrically insulated clips, especially if near high-voltage AC power lines
- AC voltmeter: assume that the potential to be measured may be hazardous initially; therefore, measure the AC voltage between the structure and ground first

4.0 PRECAUTIONS

4.1 Voltmeter

4.1.1 Two types of voltmeter are used in the CP industry, including an analog (D'Arsonval) type, as shown in Figure 4.1, and a digital type.

4.1.2 The analog instrument is very sensitive and must be treated with care during use and transport.

4.1.3 When using the analog type, attention must be given to the value of each division on the scale.

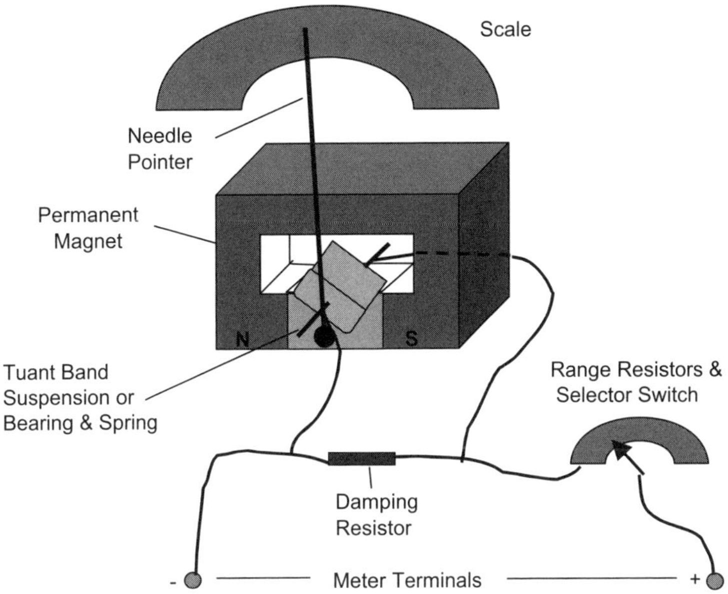

Figure 4.1 D'Arsonval-type meter movement is shown.

4.1.4 Although a dampening resistor is installed to reduce the momentum of the coil and needle, voltages exceeding the scale value will damage the instrument.

4.1.5 A digital meter is a sampling device that collects data and displays an average value. When the values are changing rapidly, the numbers keep moving, such that readings become impossible to capture. Note that the LCD readout is subject to freezing.

4.1.6 Digital voltmeters can show a reading even though not connected at all or with a poor connection or broken wire. Erroneous readings will be obtained under these conditions. With a poor connection or broken wire, the readings are seldom stable. Measuring the resistance of each meter wire while gently tugging on it will help identify a broken meter lead, and cleaning contact points will help with the connections.

4.1.7 The structure-to-electrolyte potential circuit showing each of the resistive components is given in Figure 4.2.

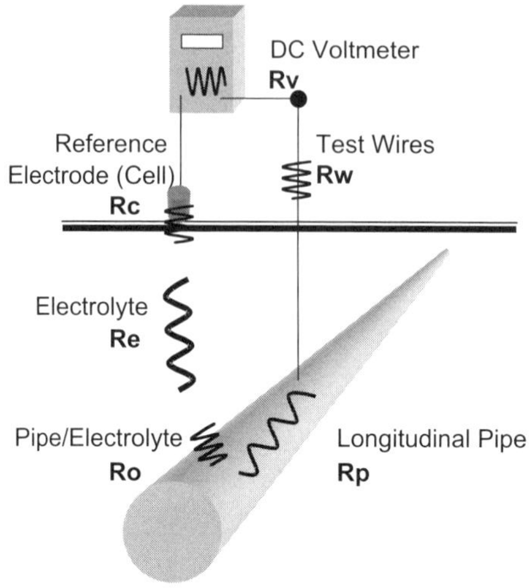

Figure 4.2 A structure-to-electrolyte potential circuit, with resistance components, is shown.

4.1.8 The internal resistance of the voltmeter (*Rv*) must be very high relative to the total resistance of the remainder of the measuring circuit (*Rw* + *Rp* + *Ro* + *Re* + *Rc*). Under most moist soil conditions, a 10-MΩ internal impedance voltmeter will be satisfactory. In conditions such as dry soil, sand, gravel, or rock, a higher internal impedance will be necessary. The addition of water to the soil may help to reduce the reference electrode contact resistance. The addition of this water must be controlled to avoid altering the environment around the structure.

4.1.9 The DC voltmeter must have AC rejection.

4.1.10 Test the voltmeter accuracy by measuring to a standard known voltage.

4.1.11 Measurement accuracy can be confirmed by the following:

4.1.11.1 The readings at two or more internal impedances are the same.

4.1.11.2 The readings at two different scales on an analog voltmeter are the same.

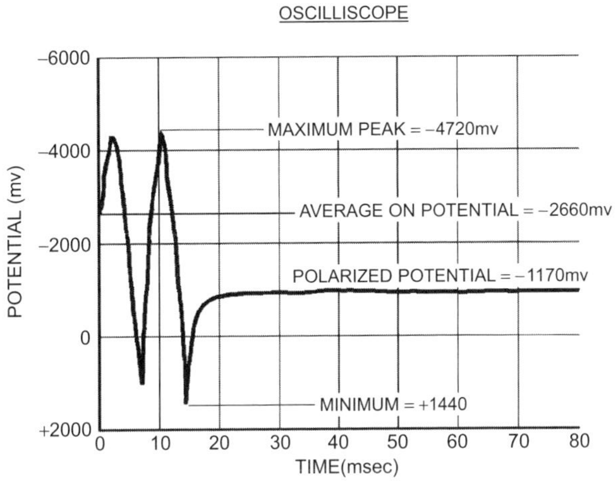

Figure 4.3 Spiking in structure-to-electrolyte potential after current interruption is shown.

4.1.12 High-speed analog-to-digital converters can record a voltage spike that may occur after the CP current interruption. The true instant OFF potential or polarized potential should be taken after this spike subsides. As the majority of these spikes are less than 0.6 second (s) (Figure 4.3), the instant OFF structure-to-electrolyte potential should be recorded between 0.6 s and 1 s after interruption.

4.2 Low Contact Resistance

Ensure that there are low contact resistances between the instrument terminals and the wires, the wire and the connectors, or the spring clip and the wire and the reference electrode.

4.3 Current-Carrying Wires

Do not use a current-carrying wire as a test lead, as shown in Figure 4.4, wire B–C. The *IR* drop in the wire will add an error to the reading, depending on the current, wire resistance, and direction of current, as illustrated by the calculations in Figure 4.4.

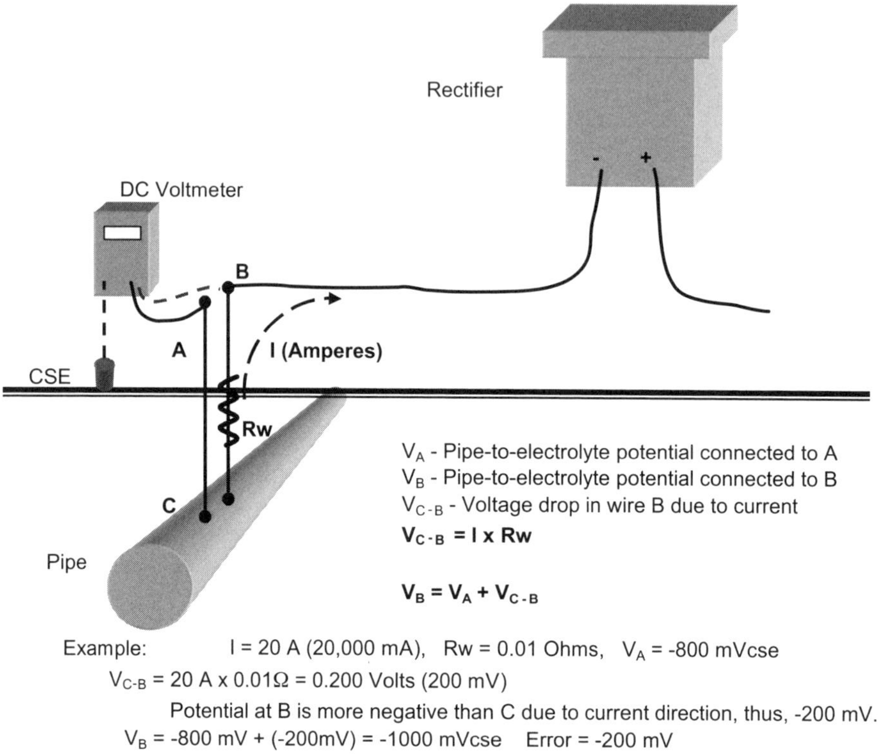

Figure 4.4 Voltage error added to (or subtracted from) potential measurement when connected to a current-carrying wire (B–C) is shown. Use only wires without current (A–C) when measuring potentials.

4.4 Reference Electrodes

4.4.1 The change in potential of the structure is desired; therefore, the reference electrode must remain stable. Routine maintenance of the reference electrode is required.[1]

4.4.2 A copper-copper sulfate (CSE) reference electrode (Figure 4.5) is normally used for structures in soil or fresh water, whereas a silver-silver chloride (SSC) reference electrode is used in high-chloride electrolytes such as seawater. Chloride contamination

1 *See* Appendix A: Reference Electrode Maintenance.

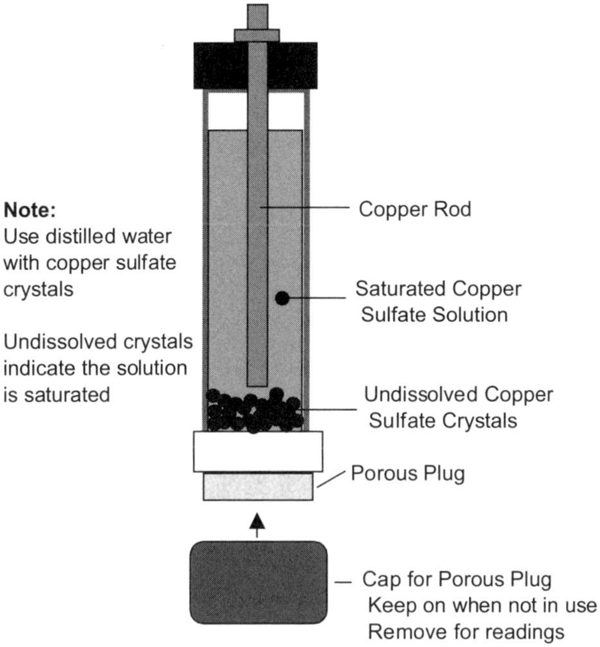

Figure 4.5 Copper sulfate reference electrode components are shown.

through the porous plug of the CSE can occur if used in high-chloride environments.

4.4.3 Other electrodes may be used, such as a saturated calomel electrode (SCE) and a hydrogen electrode; however, these are normally used in laboratory conditions.

4.4.4 A potential taken to one type of reference electrode will be different from that taken to another. A conversion from one electrode can be made when the relative potential differences between the two reference electrodes are known. An example is given in Section 6.3.

4.4.5 Reference electrodes must be clean and free of contamination. Their calibration should be confirmed daily by measuring to a standard electrode not used for other purposes.

4.4.6 Where the reference electrode is a copper-copper sulfate type, the copper rod should be of pure electrolytic copper, immersed in a cylinder containing a saturated solution of copper sulfate that is contained by a cap on one end and a porous, nonmetallic plug on the other end. Only laboratory-grade copper sulfate and distilled water are to be used for the solution. Saturation is assured by the presence of undissolved copper sulfate crystals remaining in the solution.

4.4.7 The plug must be sufficiently porous to allow an electrolytic contact between the solution and the soil or electrolyte, without allowing the copper sulfate solution to leak out.

4.4.8 Place a cap over the porous plug when not in use to keep it moist and to retain the electrode solution.

4.4.9 Remove the cap during measurements.

4.4.10 There are two types of SSC reference electrodes: an open or seawater (SW) type and a saturated type. The open type must be used in seawater; otherwise, a correction must be applied to the reading, as shown in NACE SP0176.[7.1] This correction is not required for the saturated type, which is similar in construction to the CSE, except that the electrode is silver and the solution is simulated seawater. Sections 4.4.4 and 4.4.5 also apply.

4.4.11 The reference electrode must be located near the structure.

4.4.12 If the top of the reference electrode can get wet during measurements, then encapsulate the top terminal and wire connection to ensure that it does not become part of the electrode, causing an error in the measurement.

4.4.13 Temperature has an effect on the reference electrode potential, in accordance with Equation 4.1[7.2]:

$$E_t = E_{25^\circ C} + k_t(T - 25^\circ C), \tag{4.1}$$

where

E_t	reference potential at temperature t
$E_{25^\circ C}$	reference potential at 25°C
Kt	temperature coefficient (CSE is ~0.9 mV/°C or ~0.5 mV/°F; SSC is –0.13 mV/°C or –0.07 mV/°C)

4.5 *IR* Drop

4.5.1 An *IR* drop is a voltage across a resistance, in accordance with Ohm's law ($V = IR$).

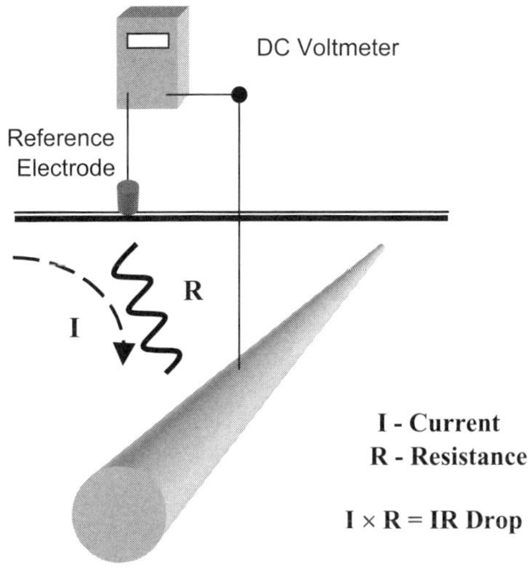

Figure 4.6 *IR* drop error between reference electrode and structure-to-electrolyte boundary is shown.

4.5.2 Although there are several *IR* drops in a structure-to-electrolyte potential circuit, the *IR* drop normally of most concern is that which occurs between the reference electrode and the structure-to-electrolyte boundary (Figure 4.6). This *IR* drop is due to a CP current in the resistance of the electrolyte and is an error in the measurement.

4.5.3 One way to reduce the *IR* drop is to bring the reference electrode close to the structure to reduce the resistance of the electrolyte. Although somewhat effective with bare structures, it is not effective with coated structures as the circuit will likely be to the nearest coating holiday, which may still be a long distance away (Figure 4.7).

4.5.4 The other approach is to bring the current to zero by interrupting all sources of current influencing that point at the same precise time (if $I = 0$ A, then $IR = 0$ V).

4.5.5 An *IR* drop between the structure contact and the reference electrode location may also occur if there is a significant distance between them (Figure 4.8) due to the current (*Ip*) and the structure resistance (*Rp*). A correction for this situation is given in Section 6.0.

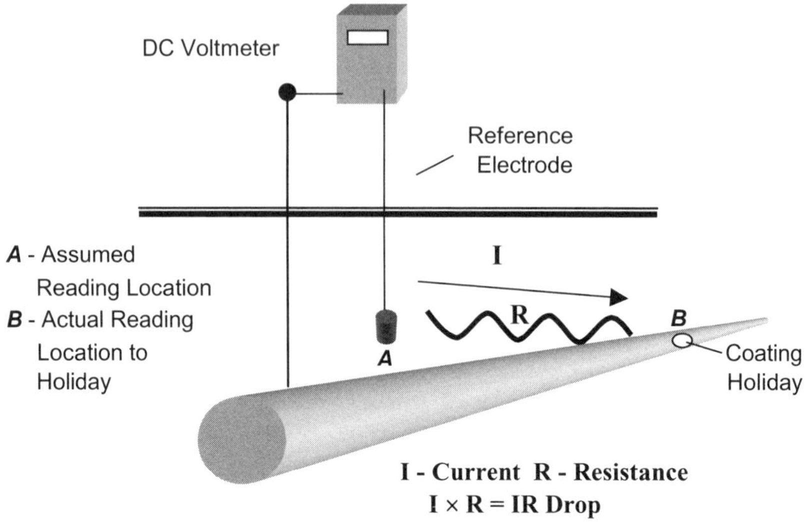

Figure 4.7 Increased *IR* drop to a remote coating holiday is shown.

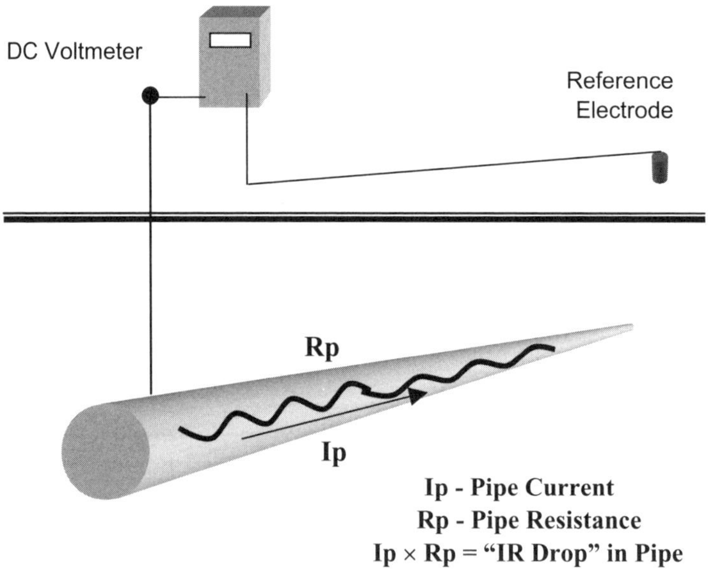

Figure 4.8 *IR* drop due to current in the pipe is shown.

5.0 PROCEDURE

5.1 Structure-to-Electrolyte Potential Equipment

5.1.1 The equipment for a structure-to-electrolyte potential measurement consists of a high-impedance voltmeter, a reference electrode, and connecting wires.

5.2 Voltmeter

5.2.1 Digital Voltmeter

5.2.1.1 Digital meters are an electronic instrument with a digital readout that is normally an LCD. Digital voltmeters may come with other features, such as AC volts, an ohmmeter, diode testing, DC amperes, DC milliamperes, and/or DC microamperes. It may also have an autoranging scale, hold, and hertz features. Special instruments or data loggers are available to store the readings, polarity, date, and time. The features of interest in a structure-to-electrolyte potential are the DC volts, autoranging scale, hold, and data logging.

5.2.1.2 Digital voltmeters without autoranging will have to be set to the appropriate scale for the measurement. After checking to ensure that there are no hazardous AC voltages present, start with the highest DC volt scale, and then reduce to the lowest scale that can capture the value being measured. The units and polarity will also be displayed.

5.2.1.3 Digital voltmeters display the units and polarity, and those with an autoranging feature will automatically change the scale, based on the value. When the measurement value exceeds a scale, the instrument will automatically change the scale and possibly the units, which may go unnoticed; for example, it may switch from a millivolt scale to a volt scale. It is, therefore, critical that the units and polarity in addition to the value be noted each time.

5.2.1.4 The hold feature allows the operator to capture a reading so that it can be recorded afterward.

5.2.1.5 The features within a data logger vary from one instrument to another and will be described in the manufacturer's operating manual. They may be programmed to enter measurements on command or at time intervals. Although the potentials may be time stamped and appear to be accurate to the second, this must be confirmed. For example, when data loggers are being synchronized, each display may show the time to the same second; however, one instrument may be at the start of the second, whereas the other may be near the end of the same second. In fact, the data of the other data logger will be recorded as if it were almost a second later. This will be evident by comparing similar wave prints of stray current potentials.

5.2.2 Analog Voltmeter

5.2.2.1 The basic components of an analog meter consist of a stationary, permanent magnet, inside of which is a moving electric coil connected to a spring or a taut band. Current passing through this coil creates an opposing magnetic field to the permanent magnet, causing the coil to rotate. A larger current in the coil will generate a greater rotation. A needle attached to this coil deflects in proportion to the coil rotation and moves along a scale to provide a continuous reading. The current in this coil comes from the measuring circuit, and the amount is dependent on the voltage across the voltmeter terminals, in accordance with Ohm's law. This is called D'Arsonval-type meter movement (Figures 4.1 and 5.1).

5.2.2.2 The scale on an analog voltmeter may be a left-hand zero or a center zero. If it is a left-hand zero, the positive of the circuit must be connected to the positive voltmeter terminal and the negative of the circuit to the negative voltmeter terminal for the needle to move to the right. Otherwise, it is pinned, and no movement will be seen. If the polarity reverses, then the voltmeter leads have to be reversed or a polarity-reversing switch changed. The needle in a zero-center voltmeter will either move upscale or downscale, depending on the polarity.

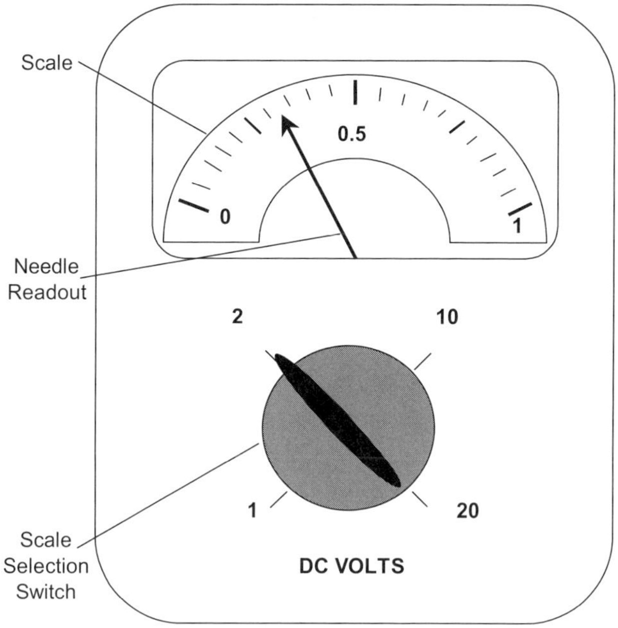

Figure 5.1 Sample of an analog meter readout is shown.

5.2.3 Reading the Analog Scale

5.2.3.1 If the analog meter has a single setting, the full-scale value will be printed on the scale.

5.2.3.2 If the meter has a multiple scale option, the full-scale value will be determined by the scale selector switch (see Figure 5.1).

5.2.3.3 Calculate the value of each major and minor division based on the number of divisions from zero to full scale.

5.2.3.4 Note where the needle is resting. If the needle has an arrowhead, then take the reading at the point of the arrow. If the needle is straight, it will have some depth. Line up the needle with the eye such that it appears to be at the thinnest point and take that reading. If there is a mirror on the scale, then line up the needle with its mirror image until there is no visible shadow in the mirror and take the reading at that point.

5.2.3.5 Calculate the reading by counting the number of divisions and multiplying by the value of the division.

5.2.3.6 Using the example in Figure 5.1, the full-scale value is 2 V_{DC}, based on the scale selector switch, which means that the scale must be multiplied by 2; that is, 0.5 will be 1.0 V_{DC}. There are 20 divisions from zero to full scale; therefore, each division is equal to 0.1 V_{DC} (2.0/20 = 0.1 V_{DC}). The needle is at 6.5 divisions from zero; therefore, the reading is (6.5 divisions × 0.1 V_{DC}/division) = 0.65 V_{DC}. In this case, the reading can be taken directly from the scale and then multiplied by the scale selection switch (0.325 [scale reading] × 2 [selector switch] = 0.65 V_{DC}), but this may not always be applicable.

5.2.3.7 Alternately, use a major division as the starting point. The major division below the needle is 0.5 V_{DC}. The needle is 1.5 divisions above this point; therefore, the reading is (0.5 V_{DC}+ [1.5 divisions × 0.1 V_{DC}/division]) = 0.65 V_{DC}.

5.3 Reference Electrode

5.3.1 Calibrate the reference electrode to a second electrode that is kept as a standard. If the measurement between the electrodes differs by more than 5 mV, then follow the servicing instructions in Appendix A. Record the results, date, and time. Calibrate the standard electrode to a laboratory calomel electrode frequently.

5.3.2 If using a digital type of voltmeter, connecting the reference electrode to the negative or "common" terminal is recommended. When connecting the reference electrode to the positive terminal, the operator must record the opposite polarity to that of the reading; that is, a positive reading must be recorded as negative as the meter leads are reversed and, thus, the polarity is reversed.

5.4 Polarity

5.4.1 The polarity of the structure-to-electrolyte potential measurement is of utmost importance. Regardless of how the voltmeter is connected, there is only one way to record the reading.

5.4.2 When referencing a structure to an electrolyte (i.e., to a reference electrode), the structure is considered positive with respect to the electrolyte. The positive terminal of the voltmeter would then be connected to the structure and the negative terminal to the reference electrode. If a measurement is taken between steel and a CSE reference electrode, the measurement will normally be negative and must be recorded as a negative value. A digital voltmeter will display a negative sign in this case. (Most digital meters will display no sign when the reading is positive.)

5.4.3 When using an analog instrument with a left-hand zero for a structure-to-electrolyte potential measurement, the connections normally have to be reversed for a deflection to the right (upscale) to occur, which is required for a reading to be seen on the scale. The structure is then connected to the negative and the reference electrode to the positive. The operator must remember, however, that the connections are reversed and that the reading is in fact still a negative value. The readings must still be recorded as being negative.

5.5 Measurement Records

5.5.1 A structure-to-electrolyte potential is obtained by measuring the difference in potential between the metal of a structure and the reference electrode with a DC voltmeter (Figure 5.2).

5.5.2 If using a digital voltmeter, connect the positive terminal of the voltmeter to the structure to ensure that there is essentially no resistance in the connection. (If an analog voltmeter is being used, the negative of the structure must be connected to the structure to obtain a measurement, but the reading is still recorded as a negative value.)

5.5.3 The other terminal of the voltmeter is connected to the reference electrode. The reference electrode porous plug cap is removed and the porous plug is placed into moist soil or water over or close to the structure. Do not immerse the opposite end of the electrode unless it has been sealed.

5.5.4 Again, a direct reading that is obtained on the meter will normally be negative and recorded as such.

5.5.5 Fluctuating readings may indicate that a connection is poor or that the wire lead has a partial break.

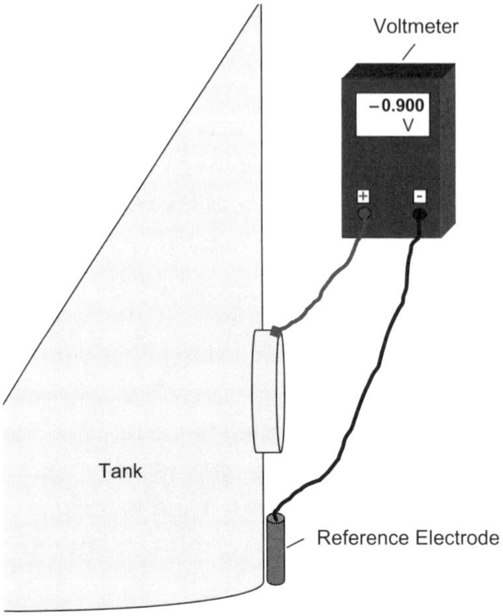

Figure 5.2 Typical structure-to-electrolyte potential is shown.

5.5.6 All measurements must include the polarity, value, units, and type of reference electrode. As an example, a negative 850 mV referenced to a CSE electrode can be recorded as $-850\ mV_{CSE}$.

5.5.7 Provided that the positive is connected to the structure and the negative to the reference electrode, a digital voltmeter will display the value, polarity, and units.

5.5.8 If the digital meter is an autoranging type, the units, and, therefore, the display, may change for each reading. Note the units each time.

5.5.9 The response of an instant OFF structure-to-electrolyte potential in an analog voltmeter will be seen as a significant hesitation in the meter needle as it moves downscale.

5.5.10 The response of a digital meter is often more difficult to predict. This is even more so if the digital meter is autoranging as the meter first decides which range to use before displaying the value. If there is a significant change from the ON and OFF potential values and the cycle is fast, then the OFF potential may

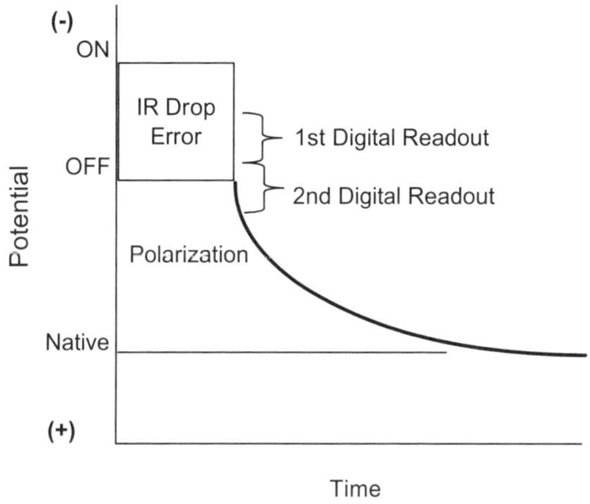

Figure 5.3 Structure-to-electrolyte response after current interruption related to criterion is shown. The criteria for cathodic protection are described in NACE SP0169.[7.3]

not be properly displayed. In this case, the autoranging should be turned off and the scale fixed to display the highest value. This is normally accomplished by pushing a range button until the desired scale is obtained.

5.5.11 The digital meter is a sampling device, and the first reading may have sampled readings during the *IR* drop portion of the interruption (Figure 5.3). For this reason, it is recommended that the second reading displayed after interruption be used as the instant OFF value.

5.6 Interrupted (ON/OFF) Structure-to-Electrolyte Potentials

5.6.1 Install synchronized current interrupters in all DC power sources and all bonds that supply current to the structure.

5.6.2 Select a long ON cycle and a short OFF to preserve as much polarization as possible. The length of the OFF cycle must be sufficient to allow the tester or the automatic recording voltmeter to capture the instant OFF value.

5.6.3 Record the length of time for the ON and OFF cycles.

5.6.4 Identify the ON and OFF readings by the length of time of the cycle. Do not assume that the most negative reading is always the ON reading because it may not be. It is imperative that reverse changes in potential be seen and recorded.

5.6.5 ON/OFF structure-to-electrolyte potentials are typically conducted before a depolarization test is completed as a baseline in addition to comparing the OFF values to the native potentials.

5.6.6 If using an automatic recording voltmeter (data logger) to record the readings, determine if there is "spiking"; otherwise, select a value that is at least 0.6 s after interruption.

5.6.7 If using a digital voltmeter, select the second reading displayed because the first reading may still have *IR* drop values as the reading is going from the ON to the OFF value (Figure 5.3).

5.6.8 If using an analog voltmeter, a noticeable reduction in the speed of the needle's descent can be seen, at which point the reading is taken as the OFF value.

5.6.9 When multiple current sources are being interrupted, the installation of a stationary data logger is recommended to observe any loss of synchronization of the interrupters. Figure 5.4 illustrates where one rectifier failed in the ON cycle and where another was out of synchronization (telluric current is also evident).

5.7 Dynamic Stray Current

5.7.1 Determine if the structure under test is affected by a dynamic DC stray current.(2)

5.7.2 Adjacent potentials that change in excess of 20% should be verified. The potential spikes should be tested immediately to confirm proper reference electrode contact with the ground by the addition of water or by exposing moist soil.

5.7.3 Telluric or other dynamic stray current activity requiring calibration can be defined as an OFF potential fluctuation exceeding 20 mV peak to peak over the duration of the testing.

---

[2] *See* Cathodic Protection Procedure No. 8: Direct Current Stray Current Interference.

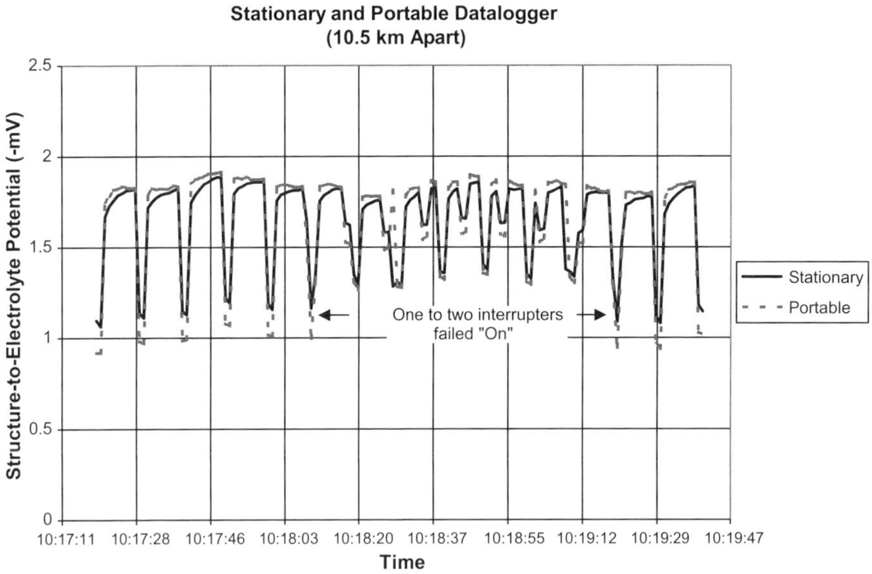

Figure 5.4 Comparison of wave prints of two data loggers 10.5 km (6.5 mi) apart, indicating current interrupter failure, in addition to some telluric activity, is shown.

5.7.4 If telluric or other dynamic stray current is detected, install data loggers near each end of the section to be tested to record structure-to-electrolyte (pipe-to-soil) potentials versus time and leave recording for a period of approximately 22–24 hours, where practical.

5.7.5 Alternately, measure structure-to-electrolyte potentials manually and record the values and the time of each reading. Plot the results to see any trends.

5.7.6 If the test section is less than 1.6 km (1.0 mi), a single data logger may be installed at the test site.

5.7.7 Record each test station structure-to-electrolyte potential with another data logger for a period of 5 min. Take close interval survey potentials as normal, except the readings are to be time stamped.

5.7.8 Figure 5.4 shows that the wave prints of a portable and stationary data logger 10.5 km (6.5 mi) apart are very similar.

Although there was a small amount of telluric current in this time period, it also demonstrated that two current interrupters failed in the ON position, typical of some Global Positioning System interrupters; therefore, an incorrect reading may have been taken manually.

5.8 Soil Coupon Tests

5.8.1 Where it is impractical to interrupt all DC power sources, a soil coupon test station can be installed to predict an *IR* drop free potential (Figure 5.5). The coupons should be sized to a typical holiday in the system. One coupon is attached to the structure with CP (CP coupon), and a second coupon is left under native conditions (native coupon). The CP coupon can be temporarily interrupted and an instant OFF potential taken. The current being impressed onto the coupon can be measured with a zero-resistance ammeter.

5.8.2 A typical installation is shown in Figure 5.5.

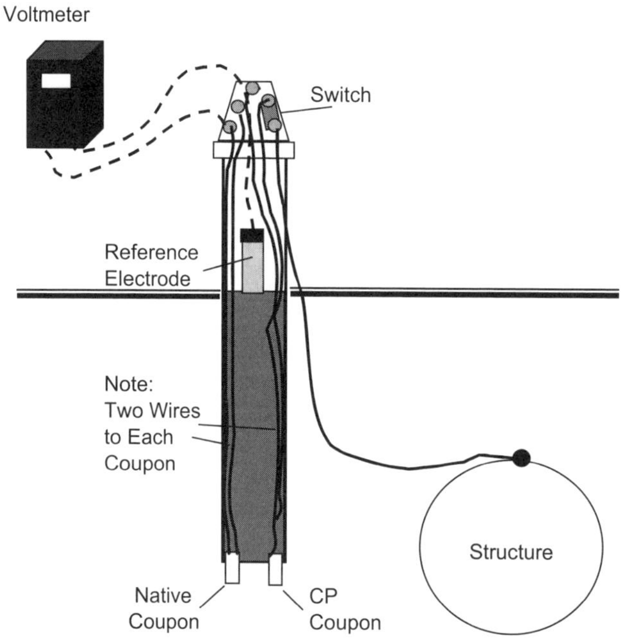

Figure 5.5 Soil coupon showing potential measurement is shown.

5.8.3 As a minimum, obtain the following potential measurements:
- On CP coupon-to-electrolyte potential
- Off CP coupon-to-electrolyte potential (switch in coupon panel turned off)
- Native coupon-to-electrolyte potential
- Polarization decay or formation on the CP coupon

5.8.4 With specialized equipment, complete the following tests:
- CP coupon current
- Linear polarization resistance corrosion rate test

6.0 ANALYSIS

The analysis of structure-to-electrolyte potentials will differ, depending on the objectives of the survey, which are covered in other modules. One main purpose for structure-to-electrolyte potentials is to determine if the criteria for CP have been met. Other uses are to confirm that the DC power source has been connected correctly, to confirm that isolating features or road casings are effectively isolated, to test for stray current interference, and to assist in troubleshooting. The special tests are covered in other CP procedure modules; however, some basic analysis is included subsequently.

6.1 Criteria for Cathodic Protection

6.1.1 CP criteria are detailed in Section 6 of NACE SP0169-2007[7.3] and in other industry-recognized codes.[7.6,7.7] [(3)] One must keep abreast of any changes to these criteria as they will be reviewed and possibly revised at regular intervals. The CP Tester must have a good understanding of the criteria as well as of the special conditions and precautionary notes that apply to these criteria, as described in this document.

6.1.2 There are currently three structure-to-electrolyte potential criteria for submerged or buried steel structures in the absence of specific data that demonstrate that adequate CP has been applied, as illustrated in Figure 6.1. These include the following:

6.1.2.1 "A negative (cathodic) potential of at least 850 mV with the cathodic protection applied" (with respect to a CSE

[3] Use most recent version. ISO 15589-1[7.7] includes only criteria in 6.1.2.2 and 6.1.2.3.

contacting the electrolyte). *IR* drops between the reference electrode and the structure-to-electrolyte boundary are an error in this reading and must be removed before applying this criterion, as illustrated by Equation (6.1). NACE SP0169[7.3] discusses methods to evaluate the *IR* drop, which is essentially the difference between the ON potential and the instant OFF potential:

$$E_c = (E_{\text{on}} - IR), \tag{6.1}$$

where

- E_c potential for criterion ($-850\ \text{mV}_{\text{CSE}}$ or more negative)
- E_{on} potential with current applied
- IR voltage drop between the reference electrode and the structure-to-electrolyte boundary

6.1.2.2 "A negative polarized potential of at least 850 mV relative to a saturated copper-copper sulfate reference electrode contacting the electrolyte." This can be obtained by interrupting all current sources influencing the structure and taking an instant OFF potential.

6.1.2.3 "A minimum of 100 mV of cathodic polarization between the structure surface and a stable reference electrode contacting the electrolyte." Polarization is the change in potential from the native or free corroding potential and the instant OFF potential, as illustrated by Equation (6.2). It can be measured during formation or decay:

$$E_p = (E_{\text{off}} - E_{\text{native}}), \tag{6.2}$$

where

- E_p polarization for criterion (100 mV or greater)
- E_{off} potential with all current momentarily interrupted
- E_{native} native potential before CP current is applied or depolarized potential after current has been left off

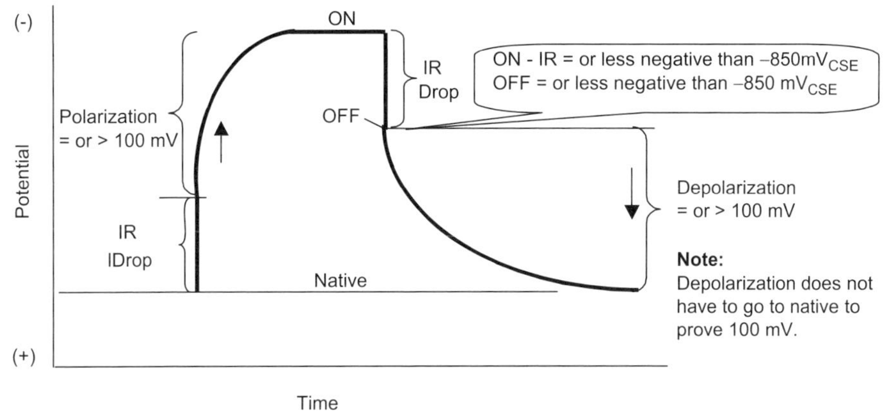

Figure 6.1 Illustration of CP criterion for steel is shown.

6.1.3 Figure 6.1 illustrates a current being applied on the left side and then interrupted on the right side of the profile. The *IR* drop is seen immediately on applying the current and immediately after interruption. The remaining change is due to polarization. Note that the first two criteria arrive at the same point (-850 mV$_{CSE}$) using different techniques. The 100-mV criterion can be seen during formation or decay in this figure.

6.1.4 Only one of these criteria needs to be met. For example, if a polarized potential is less negative than -850 mV$_{CSE}$, a depolarization survey may prove that the 100-mV criterion is being achieved.

6.1.5 Where two or more metals are coupled, the 100-mV polarization cannot be used, unless the potential of the most active (most electronegative) metal is known.

6.1.6 Other metals have different criteria and are given in NACE SP0169.[7.3]

6.2 *IR* Drop Considerations

6.2.1 *IR* drop consideration means the application of sound engineering practice to determine the significance of *IR* drop error included in the structure-to-electrolyte potential between the reference electrode and the structure and, in certain cases, in the structure itself.

6.2.2 Where practical, the *IR* drop is removed by taking an instant OFF structure-to-electrolyte potential. If the current (I) can be brought to zero, then the *IR* drop is also zero.

6.2.3 If the ON potential at a given location remains the same, the current applied is the same, the potentials are taken under similar conditions, and there has been no change to the structure, then the *IR* drop as determined by the last ON/OFF survey can be used for the same location.

6.2.4 Note that a fixed *IR* drop should not be applied to all test locations, unless there has been sufficient testing to prove that it is a realistic value.

6.2.5 If it can be shown that there is a lack of corrosion at the past and present structure-to-electrolyte potentials, the ON potential including the *IR* drop can be accepted.

6.2.6 When using a soil coupon, compare the instant OFF coupon-to-electrolyte potential to the polarized potential criterion.[7.3] Determine the polarization by comparing the OFF coupon-to-electrolyte potential to the native potential for the amount of polarization.

6.3 Reference Electrode Conversion

6.3.1 Where more than one type of reference electrode was used, it is desirable to convert them all to the same reference, normally CSE. A calibration of one electrode to the other should be made for this conversion.

6.3.2 An example of conversion is illustrated in Figure 6.2, using the calibration of Ag-AgCl (SCE) referenced to CSE as −50 mV, SCE with respect to CSE as −70 mV, and zinc referenced to CSE as −1100 mV. These conversion factors differ somewhat in the literature, and the actual amount should be determined by measurement.[7.1–7.5] In this case, +250 mV_{Zn} converts to −850 mV CSE, and −850 mV_{CSE} converts to −800 mV_{SCE}.

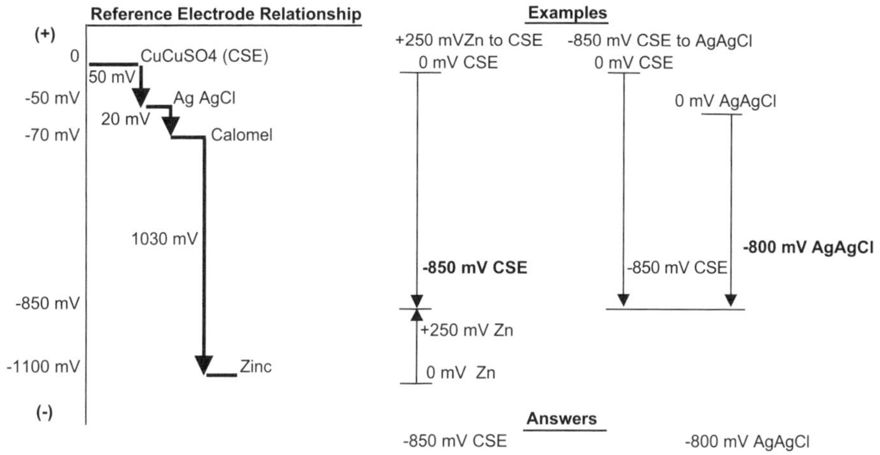

Figure 6.2 Reference electrode relationships and conversion examples are shown.

Note that the calibration differences will change with temperature (see Section 4.4.13).

6.4 Direct Current Power Source Interruption

6.4.1 Review stationary data log profiles to confirm that the interrupters at all influencing rectifiers continued to function during the test (see Figure 5.4).

6.4.2 Identify any structure-to-electrolyte potential readings that may have been affected by an interrupter malfunction.

6.4.3 Note amount of depolarization that took place during the interruption period.

6.5 Dynamic Stray Current

6.5.1 Telluric or other dynamic stray current activity requiring calibration is defined as an OFF potential fluctuation exceeding 20 mV peak to peak over the duration of the testing.

6.5.2 The true potential at the stationary data loggers should first be determined by either a quiet period or an average over the test period. For each potential measurement along the line, the difference in potential between that and the stationary data logger at the same moment in time is to be determined. This difference added to the difference to the true potential at the stationary data logger is the correction factor for the portable data logger reading. Equations are presented for the following situations:

- If two stationary data loggers are used, the following method, in Equations (6.3) and (6.4), apply:

$$\acute{\varepsilon}b = [\acute{\varepsilon}a(c - b)/c] + [\acute{\varepsilon}c(b - a)/c], \tag{6.3}$$

where

a	first stationary potential location
b	portable potential location
c	second stationary potential location
$\acute{\varepsilon}a$	error in potential at a at time x
$\acute{\varepsilon}b$	error in potential at b at time x
$\acute{\varepsilon}c$	error in potential at c at time x

and

$$Ep = Ep_{\text{measured}} - \acute{\varepsilon}b, \tag{6.4}$$

where

Ep	true potential at the portable data logger location
Ep_{measured}	potential at the portable data logger location

- If only one stationary data logger was used, then Equation (6.5) applies:

$$Ep = Es - \Delta(Esa - Epa), \tag{6.5}$$

where

Ep	true potential at the portable data logger location
Es	true potential at the stationary location
Esa	stationary potential at time a during the data logging
Epa	portable potential at time a during the data logging

6.5.3 Other methods can also be used to correct for dynamic stray currents.

6.5.4 The stationary recorder will also serve to verify synchronized interruption and if depolarization occurred during the survey interruption period.

7.0 REFERENCES

7.1 NACE Standard SP0176-2003, "Corrosion Control of Submerged Areas of Permanently Installed Steel Offshore Structures Associated with Petroleum Production" (Houston, Texas: NACE International, 2003), Figure 1.

7.2 F.J. Ansuini, J.R. Dimond, "Factors Affecting the Accuracy of Reference Electrodes," *MP* Vol. 33, No. 11 (1994): pp. 14–17.

7.3 NACE Standard SP0169-2007, "Control of External Corrosion on Underground or Submerged Metallic Piping Systems" (Houston, Texas: NACE International, 2002).

7.4 W. von Baekmann, W. Schwenk, W. Prinz, *Handbook of Cathodic Corrosion Protection*, 3rd ed. (Houston, Texas: Gulf Publishing Company, 1997), p. 80.

7.5 M.H. Peterson, R.E. Grover, *Tests Indicate the Ag/AgCl Electrode Is Ideal Reference Cell in Sea Water*, *MP* Vol. 11 (1972): pp. 19–22.

7.6 CGA Recommended Practice OCC-1-2005, *Control of Corrosion on Buried or Submerged Metallic Piping Systems*, (Ottawa, Ontario: Canadian Gas Association, 2005).

7.7 International Standard, ISO 15589-1, *Petroleum and natural gas industries – Cathodic protection of pipeline transportation systems – Part 1: On-land pipelines*, (International Standards Organization, Web: *www.iso.org*. Published in Switzerland, 2003).

APPENDIX A
REFERENCE ELECTRODE MAINTENANCE

1.0 COPPER-COPPER SULFATE REFERENCE ELECTRODE (CSE)

1.1 General

A CSE consists of a pure copper rod and a saturated copper sulfate solution, contained by a plastic tube with a porous plug on one end and a plastic cap holding the rod at the other end. The plastic tube is opaque but may have a vertical window to observe the level of copper sulfate solution and crystals (see Figure A.1).

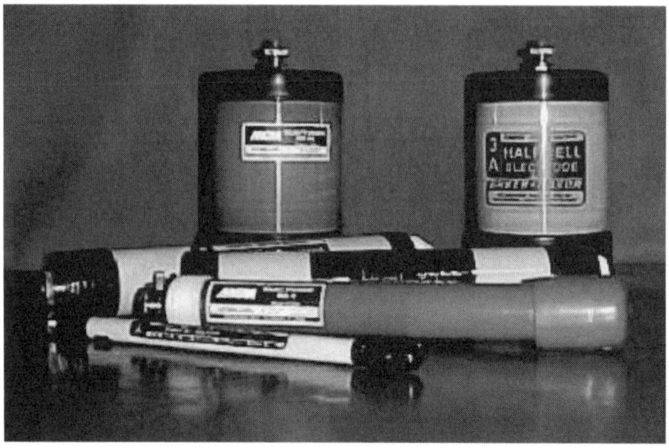

Figure A.1 CSE reference electrodes are shown. Photo from NACE CP Technician Course.

The normal use of a CSE is in fresh water or soils. The copper sulfate solution must be saturated, and this can easily be confirmed by the fact that not all crystals are dissolved. When crystals are no longer present, copper sulfate crystals should be added until some remain that are not dissolved.

A CSE can become contaminated by the addition of other metal ions such as iron, either with the water or by cleaning the copper rod with a metal abrasive. It can also be contaminated when used in a high-chloride environment because the chloride ions can migrate through the porous plug.

The reference electrode(s) should be calibrated every day before use.

1.2 Calibration

1.2.1 Measure the potential difference to a standard reference electrode. In the field, a standard reference electrode is a CSE that has not been used other than for calibration. Preferably, the standard CSE should be calibrated to a laboratory calomel electrode.

1.2.2 Place the electrodes together in a plastic container of chloride-free water (Figure A.2). Measure the potential difference between the field CSE and the standard CSE. A potential difference of 5 mV or less is required to continue using the field CSE. If the difference in potential is greater than 5 mV, the CSE must be serviced before using again.

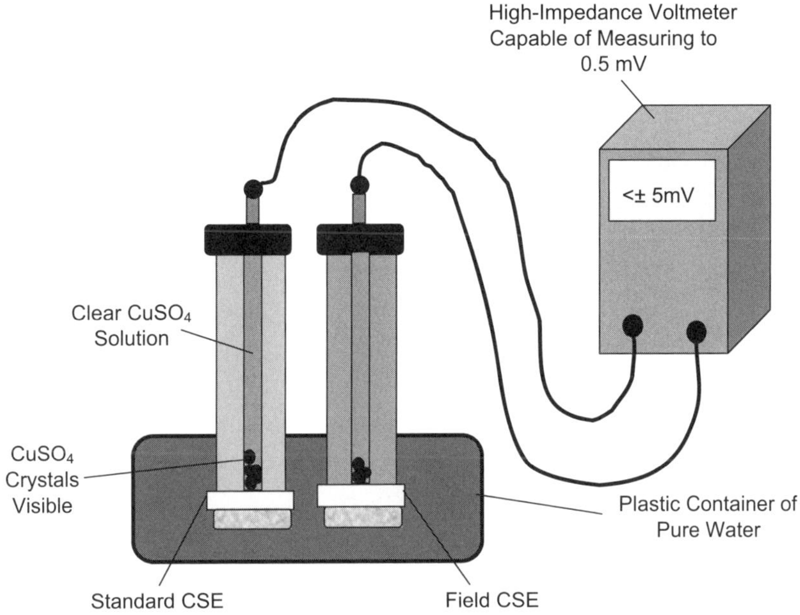

Figure A.2 Reference electrode calibration is shown.

1.3 Servicing a Copper-Copper Sulfate Reference Electrode

1.3.1 Empty the CSE and properly dispose of the contents.

1.3.2 Remove the copper rod/cap and porous plug.

1.3.3 Flush the tube with distilled or deionized water.

1.3.4 Soak the porous plug and flush with distilled or deionized water.

1.3.5 With a nonmetallic abrasive, such as silica sandpaper, clean the copper rod to a bright surface.

1.3.6 Inspect the O-rings and replace if damaged.

1.3.7 Thread the porous plug onto the end of the tube. (Alternately, install the copper rod, leaving one end open to fill.)

1.3.8 Add copper sulfate crystals to a maximum of one-fourth the volume of the tube.

1.3.9 Add distilled or deionized water and confirm that some copper sulfate crystals remain undissolved. If not, add more crystals until some crystals remain. (Alternately, a saturated copper sulfate solution can be premixed in a clean, nonmetallic container. In this case, fewer copper sulfate crystals are required in the electrode chamber.)

1.3.10 Install the copper rod (porous plug) and ensure that both ends are sealed properly.

1.3.11 Shake the solution and confirm that it is ~95% full and that some copper sulfate crystals are still evident.

1.3.12 Repeat the calibration procedure in Appendix A, Section 1.2.

1.3.13 If the calibration is still more than 5 mV, then disassemble and confirm that the rod does not have embedded particles.

1.3.14 Confirm that the water is distilled or deionized and that the copper sulfate crystals have not become contaminated.

1.3.15 Repeat the preceding servicing procedure.

1.3.16 If the calibration cannot be achieved by successive servicing, the field CSE must be replaced.

2.0 SILVER-SILVER CHLORIDE REFERENCE ELECTRODE FOR USE IN SEAWATER

2.1 General

The SCE electrode for use in seawater (SCE_{SW}) reference electrode requires seawater as the electrolytic couple. If used in water of higher resistivity than seawater, a correction must be applied to the reading in accordance with NACE SP0176.[7.1]

2.2 Calibration

2.2.1 Measure the potential difference to a standard SSC_{SW} reference electrode. A standard reference electrode is one that has not been used other than for calibration.

2.2.2 Place the electrodes together in a plastic container of seawater or manufactured seawater similar to that shown in Figure A.2. Measure the potential difference between the field SSC_{SW} and the standard SSC_{SW}. A potential difference of 5 mV or less is required to continue using the field SSC_{SW}. If the difference in potential is greater than 5 mV, the field SSC_{SW} must be serviced before using again.

2.3 Servicing a Silver-Silver Chloride Reference Electrode for Use in Seawater

2.3.1 Remove the weight and perforated tube protecting the silver electrode.

2.3.2 Carefully clean the silver electrode with a clean, soft cloth.

2.3.3 Reassemble and recalibrate in accordance with Appendix A, Section 2.2.

2.3.4 If the calibration cannot be achieved, discontinue its use and either return to the manufacturer or dispose.

3.0 CLOSED SILVER-SILVER CHLORIDE REFERENCE ELECTRODE

3.1 General

The closed SSC (SSC_C) consists of a silver rod coated with silver chloride in a manufactured seawater solution, contained by a plastic tube with a porous plug on one end and a plastic cap holding the rod at the other end. The plastic tube is opaque but may have a vertical window to observe the level of solution.

3.2 Calibration

3.2.1 Measure the potential difference to a standard SSC_C reference electrode. A standard reference electrode is one that has not been used other than for calibration.

3.2.2 Place the electrodes together in a plastic container partially filled with water, similar to that shown in Figure A.2. Measure the potential difference between the field SSC_C and the standard SSC_C. A potential difference of 5 mV or less is required to continue using the field SSC_C. If the difference in potential is greater than 5 mV, the field SSC_C must be serviced before using again.

3.3 Servicing a Closed Silver-Silver Chloride Reference Electrode

3.3.1 Empty the SSC_C and properly dispose of the contents.

3.3.2 Remove the silver rod/cap and porous plug.

3.3.3 Flush the tube with distilled or deionized water.

3.3.4 Soak the porous plug and flush with distilled or deionized water.

3.3.5 With a clean, soft cloth, clean the silver rod.

3.3.6 Inspect the O-rings and replace if damaged.

3.3.7 Thread the porous plug onto the end of the tube. (Alternately, install the silver rod, leaving one end open to fill.)

3.3.8 Add manufactured seawater.

3.3.9 Install the silver rod (porous plug) and ensure that both ends are sealed properly.

3.3.10 Confirm that the tube is ~95% full.

3.3.11 Repeat the calibration procedure in Appendix A, Section 3.2.

3.3.12 If the calibration is still more than 5 mV, then disassemble and confirm that the rod does not have embedded particles.

3.3.13 Confirm that the solution is a seawater equivalent.

3.3.14 Repeat the preceding servicing procedure.

3.3.15 If the calibration cannot be achieved by successive servicing, the field SCE_C must be replaced.

CATHODIC PROTECTION PROCEDURE NO. 3

Direct Current Measurements 3

1.0 INTRODUCTION

The purpose of this procedure is to measure direct current (DC) in an electrical circuit related to cathodic protection (CP).

DC measurements may be made by an ammeter or a clamp-on ammeter or calculated by measuring a millivolt drop across a shunt or known resistance. In CP, current measurements are needed for the output of a DC power source, continuity bond current, interference bond current, and to complete various tests.

Persons qualified to take DC measurements include CP Testers, CP Technicians, CP Technologists, and CP Specialists or equivalent.

Complete a hazard assessment before taking DC measurements. The circuit must be turned off and locked out/tagged out before installing an ammeter.

2.0 TOOLS AND EQUIPMENT

The following equipment is needed, in whole or in part, depending on the current test:

- DC ammeter
- DC clamp-on ammeter
- Optional DC clamp-on probe added on to DC voltmeter

- Voltmeter:
  - High-input impedance DC voltmeter to measure from 0 to 4,000 mV
  - Optional DC voltmeter/data logger with a memory programmed to store value, units, polarity, time, and date, either on command or at time intervals
  - Alternating current (AC) voltage circuit
- Shunt:
  - Current rating greater than expected current value
  - Low resistance to minimize influence on circuit
- Test leads, complete with electrically insulated spring clips or connectors
- Wire reel
- Regulated battery supply or 6- to 12-V battery and variable 1.5-Ω 100-W resistor

3.0 SAFETY EQUIPMENT

- Standard safety equipment, as required by the company's safety manual and regulations
- Lock-out/tag-out kit
- Electrically insulated meter probes and clips
- Assume that the potential to be measured may be hazardous initially; therefore, measure the AC voltages first, especially near AC power lines

4.0 PRECAUTIONS

4.1 Current Direction

Conventional current direction through a conductor is from positive (+) to negative (−). A negative indication on a digital meter indicates that the meter leads are connected opposite to the actual direction of current; that is, the negative meter lead is actually the positive side of the circuit.

4.2 Ammeter

An ammeter must be inserted in an electrical circuit; therefore, make certain that the circuit has been turned off and locked out/tagged out before installing. The ammeter adds resistance to the circuit and will, therefore, cause a reduction in current from the current before it was installed.

4.3 Direct Current Clamp-On Ammeter[7.1]

A DC ammeter operates on an electromagnetic principle, measuring the electromagnetic field caused by a current in a conductor. The coil must be clamped around only one conductor as the return conductor nullifies the magnetic field of the first. The clamp must be centered on the single conductor for the most accurate measurement. Note the polarity on the coil, reverse it, and take a second reading to confirm the value.

4.4 Voltmeter

The voltmeter is used to measure the voltage drop across the shunt due to the current and the shunt resistance (Ohm's law, $V = I \times R$). The internal resistance of the voltmeter must be high to avoid drawing too much current from the shunt circuit.

4.4 Shunt

A shunt is an accurately calibrated, low-resistance resistor that may be rated in ohms or in millivolts and amperes (e.g., 50 mV, 10 A). Each shunt has a current-carrying connection and a separate connection for a voltage reading. The current-carrying connection must *not* be used for voltage measurements as error can be introduced into the measurement.

4.5 Pipeline Current

A current in a pipeline can be measured either by using a clamp-on ammeter or by calibrating a portion of the pipeline as a current shunt, often called a "current span."

4.5.1 Direct Current Pipe Clamp-On Ammeter

A coil with a clamp that is sized for the diameter of the pipeline must be used. The coil must be clamped properly and centered on the pipe. When taking the measurement, note both the scale setting and multiplier on the instrument and the polarity on the coil before recording the value. Repeat the measurement with the coil in the reverse direction.

4.5.2 Pipeline Current Span

The current span is defined by test leads attached to the pipe spaced far enough apart so that the current in the resistance of the pipe span will cause a measurable voltage drop.

4.5.2.1 A single test wire on each end requires that pipe resistance tables determine the resistance of the pipe span. This means that the distance between the test leads, the pipe diameter, and the wall thickness or pipe weight must be known and that the pipe wall thickness must remain constant, without any other attachments.

4.5.2.2 An alternate method is to install two test wires at each end, in which case the current span can be calibrated as described in Section 5.4. In this case, the pipe span should be calibrated near the current values that are expected in a later measurement.

5.0 PROCEDURE

5.1 Direct Current Ammeter

5.1.1 De-energize the circuit, lock out/tag out, and confirm that it is deenergized before installing an ammeter. The ammeter is to be inserted in the circuit with the positive terminal toward the positive of the power source and the negative terminal toward the negative of the power source. If the reading is negative, confirm that there is not another power source.

5.1.2 Set to the highest ampere scale, remove the lock-out/tag-out when safe, and energize.

5.1.3 Reduce the scale such that the reading on two scales is similar. Note that the internal resistance of the ammeter increases as the scale is reduced. This, in turn, reduces the current in the circuit as the ammeter resistance is in series with the electrical circuit, which may or may not be significant.

5.1.4 Record the current direction, value, and units (amperes or milliamperes).

5.2 Direct Current Clamp-On Ammeter

5.2.1 The DC clamp-on ammeter may be self-contained, with a built-in readout, or it may be an add-on to a DC millivoltmeter. In each case, it is important to read the manufacturer's directions on its operation and to zero the meter before taking a measurement. A multiplier is required to calculate the current in an add-on clamp.

5.2.2 If there are no exposed electrical terminals near the point of measurement, the power source need not be turned off, but if there are exposed electrical terminals, turn off and lock out/tag out the AC disconnect while attaching the ammeter.

5.2.3 Set to the highest ampere scale, if not an autoranging type.

5.2.4 Clamp the coil around a single conductor, with the positive side of the coil toward the positive of the power source. If the reading is not positive, then confirm that there is not another power source.

5.2.5 Adjust the scale to the closest range for the measurement.

5.2.6 Reverse the coil and confirm that the values are similar. An average of the two readings is often close to the true value.

5.2.7 Record the current direction, value, and units (amperes or milliamperes).

5.3 Shunt

5.3.1 A shunt is an accurately calibrated, low-resistance device. The principle is to measure a voltage across the resistance of the shunt and then calculate the current. An ammeter is, in fact, a voltmeter that measures a millivolt drop across a shunt, but the scale of this meter is calibrated such that it will read in amperes. In some cases, the shunt is external to the ammeter, and in other cases, the shunt is installed inside the ammeter enclosure.

5.3.2 Shunts come in different forms, as shown in Figure 5.1, but in all cases, the material between two specific contacts is the portion that is calibrated. In the case of the wire shunt, or the lower left shunt in Figure 5.1, the millivolt readings must be taken between the two protruding pins. The inside small bolts on the remaining shunts are the contacts for the millivolt reading.

5.3.3 A shunt must be installed in series in the electrical circuit, and, thus, power to the circuit must be turned off and locked

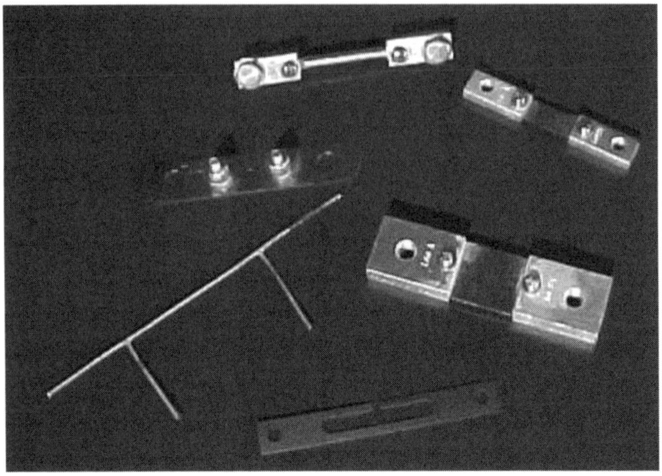

Figure 5.1 Types of shunts are shown. (Measurement is to be taken between inside small bolts or between protruding pins [not current-carrying bolts].)

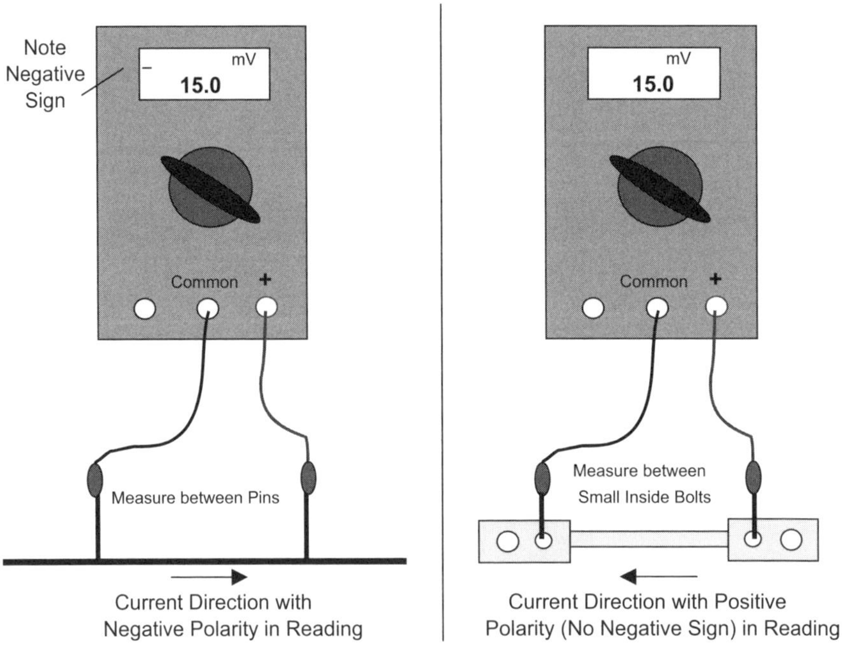

Figure 5.2 Voltmeter connection for shunts is shown.

out/tagged out before installation. Often, shunts remain in the circuit once installed. In this way, the resistance of the shunt in series during a measurement does not alter the resistance of the circuit.

5.3.4 The millivolt drop across the shunt is first measured with a voltmeter, either across two wires extending from the shunt or across the two smaller bolts in the block at each end of the shunt (Figure 5.2). The direction of current with two different polarities displayed is also shown in Figure 5.2.

5.3.5 Note that the negative polarity in the left meter in Figure 5.2 indicates that the meter leads are connected in reverse; that is, the positive lead is actually connected to the negative side of the shunt, and the negative (common) terminal is actually connected to the positive side of the shunt.

5.3.6 The current is then calculated as described in Section 6.3.

5.4 Pipeline Current Span[7.1,7.2]

5.4.1 Two-Wire Span

5.4.1.1 Determine the pipe outside diameter and wall thickness.

5.4.1.2 Determine the distance between the two wires and confirm that the pipe wall is uniform between the wires. Also confirm that there are no valves, pipeline fittings, or taps in between.

5.4.1.3 Measure the millivolt drop between the two wires and calculate the current, as described in Section 6.4.

5.4.2 Four-Wire Span

If the resistance or calibration factor of the pipe span is not known, determine it with the following test:

5.4.2.1 Impress a known amount of current between the two outside wires, as shown in Figure 5.3. The two-pole, double-throw switch to reverse polarity is optional as the leads themselves can be reversed.

5.4.2.2 Measure the millivolt drop between the inside test wires before current is applied and note the polarity, value, and units (mV_{off}).

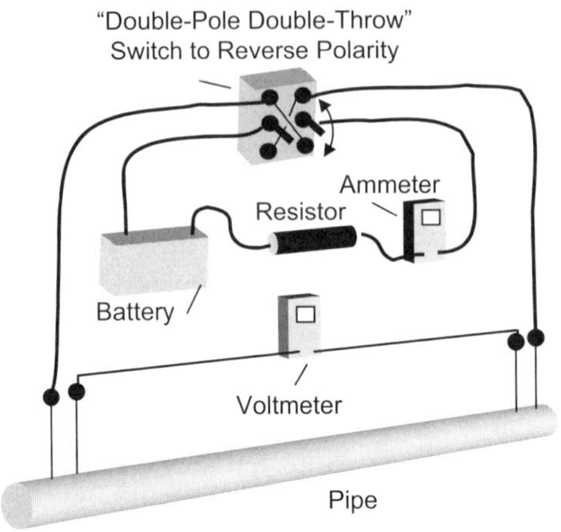

Figure 5.3 Calibration of four-wire pipe current span is shown.

5.4.2.3 Apply a current near the value of current expected to be in the pipeline and record the polarity, value, and units (I_{on}).

5.4.2.4 Measure the millivolt drop across the inside test wires with the current applied (mV_{on}).

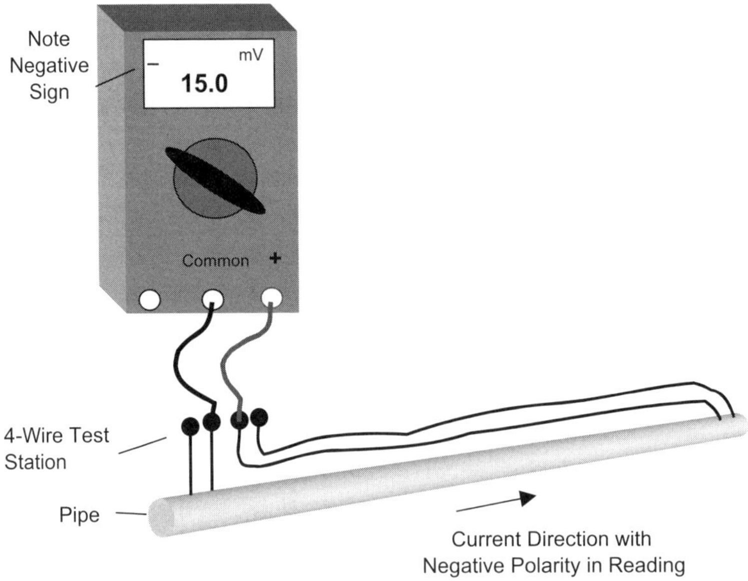

Figure 5.4 Voltmeter connection for a four-wire pipe current span is shown.

5.4.2.5 Reverse the direction of current and repeat the current and millivolt measurements.

5.4.3 To determine the current in the pipeline, connect the voltmeter between the two inside wires and measure the millivolt drop for all future pipe current readings, as shown in Figure 5.4.

5.4.3.1 Record the polarity, value, and units as well as the polarity of the meter connection to the wire color and terminal number.

5.4.3.2 Calculate the current in the pipeline, as described in Section 6.4.

6.0 ANALYSIS

6.1 Direct Current Ammeter

6.1.1 The DC ammeter will either give a direct reading in amperes (milliamperes), or there will be a scale/multiplier to apply to the meter reading.

6.1.2 Note the polarity, value, and units (e.g., +1.5 A).

6.2 Direct Current Clamp-On Ammeter

6.2.1 A self-contained DC clamp-on ammeter will either give a direct reading in amperes (milliamperes), or a scale multiplier may have to be applied to the value reading.

6.2.2 The add-on types of clamp-on probe to a DC voltmeter will read out in millivolts and will have a calibration factor in amperes per millivolt. Convert the millivolt measurement to amperes using Equation (6.1):

$$I = \text{mV}_{\text{measured}} \times \text{CF}_p, \tag{6.1}$$

where

I	actual current (amperes)
$\text{mV}_{\text{measured}}$	voltage measured across shunt (millivolts)
CF_p	clamp-on probe calibration factor (amperes per millivolt)

6.2.3 The polarity is determined by a combination of the polarity of the reading and the orientation of the coil. If the polarity of the reading is positive, the current is in the direction from positive to negative on the coil. A negative reading indicates that the current is in the opposite direction.

6.3 Shunt

The current can be calculated in one of three ways, including a ratio method, a calibration factor, or a resistance method using Ohm's law. These methods are discussed subsequently, with examples.

6.3.1 Shunt Ratio Method

The current in a shunt will yield a voltage drop (millivolts) across the shunt that is in proportion to the shunt rating. The current can then be calculated using Equation (6.2):

$$I = \frac{\mathrm{mV}_{\mathrm{measured}}}{\mathrm{mV}_{\mathrm{rated}}} \times I_{\mathrm{rated}}, \tag{6.2}$$

where

I	actual current (amperes)
$\mathrm{mV}_{\mathrm{measured}}$	voltage measured across shunt (millivolts)
$\mathrm{mV}_{\mathrm{rated}}$	voltage rating of shunt (millivolts)
I_{rated}	current rating of shunt (amperes)

Note that the specified units must be used (i.e., millivolts and amperes).

If 20 mV is measured across a 50-mV, 10-A shunt, then the current in the shunt is

$$\begin{aligned} I &= \frac{20\,\mathrm{mV}}{50\,\mathrm{mV}} \times 10\,\mathrm{A} \\ &= 4.0\,\mathrm{A}. \end{aligned}$$

6.3.2 Shunt Calibration Factor

A calibration factor for shunts commonly used in the CP industry is given in Table 6.1, or it can be calculated from the shunt rating in amperes per millivolt, as shown in Equation (6.3):

$$SF = \frac{I_{rated}}{mV_{rated}}, \tag{6.3}$$

where

SF	shunt calibration factor (amperes per millivolt)
I_{rated}	shunt current rating (amperes)
mV_{rated}	shunt voltage rating (millivolts)

The voltage measured across the shunt in millivolts is then multiplied by the calibration factor to calculate the current (Equation [6.4]):

$$I = mV_{measured} \times SF, \tag{6.4}$$

where

I	actual current (amperes)
$mV_{measured}$	voltage measured across shunt (millivolts)
SF	shunt calibration factor (amperes per millivolt)

For example, the calibration factor for a shunt (SF) with a 50-mV, 10-A rating can be taken from Table 6.1 or calculated by dividing the rated current (amperes) by the rated voltage (millivolts):

$$SF = \frac{10\,A}{50\,mV} = 0.2\,A/mV.$$

If 20 mV is measured, then the current is calculated using Equation (6.4):

$$\begin{aligned} I &= 20\,mV \times 0.2\,A/mV \\ &= 4.0\,A. \end{aligned}$$

6.3.3 Shunt Resistance Method

Ohm's law can be applied to determine the current in a shunt. Resistances of shunts commonly used in CP are given in Table 6.1. The first two shunts in this table may not have any description on them, but when purchased, a label identifying the

Table 6.1 Shunt Resistance and Factors

Shunt Rating		Shunt Resistance,	Shunt Factor,
Millivolts	Amperes	$R_S(\Omega)$	SF (A/mV)
–	–	0.01	0.1
–	–	0.001	1.0
50	1	0.05	0.02
50	2	0.025	0.04
50	3	0.0167	0.06
50	4	0.0125	0.08
50	5	0.01	0.1
50	10	0.005	0.2
50	15	0.033	0.3
50	20	0.0025	0.4
50	25	0.002	0.5
50	30	0.00167	0.6
50	40	0.00125	0.8
50	50	0.001	1.0
50	60	0.0083	1.2
50	75	0.00067	1.5
50	100	0.0005	2.0
100	100	0.001	1.0

resistance should be attached to them or to the panel, which must be kept.

The current in the shunt is then calculated by another form of Ohm's law, as shown in Equation (6.5):

$$I = \frac{V_{measured}}{R_S}, \tag{6.5}$$

where

I	current in shunt (amperes)
$V_{measured}$	voltage measured across shunt (volts)
R_S	resistance rating of shunt (ohms)

Note that the shunt voltage must be converted from millivolts to volts when using resistance.

If the resistance of the shunt (R_S) is not known but the rating in millivolts and amperes is known, the resistance can be calculated using Equation (6.6):

$$R_S = \frac{V_r}{I_r}, \tag{6.6}$$

where

R_s	rated resistance of the shunt (ohms)
V_r	voltage rating of shunt (volts)
I_r	current rating of shunt (amperes)

Note again that the shunt voltage rating must be converted from millivolts to volts with the current in amperes to calculate the resistance in ohms.

Using the example given previously of 20 mV measured across a 50-mV, 10-A shunt, the rated shunt resistance is

$$\begin{aligned} R_S &= \frac{0.050\ \text{V}}{10\ \text{A}} \\ &= 0.005\ \Omega \end{aligned}$$

and

$$\begin{aligned} I &= \frac{0.020\ \text{V}}{0.005\ \Omega} \\ &= 4.0\ \text{A}. \end{aligned}$$

Any of the three methods can be used to determine the current in a shunt.

6.4 Pipeline Current Span[7.1,7.2]

6.4.1 Two-Wire Span

6.4.1.1 Confirm the pipe diameter and wall thickness.

6.4.1.2 Confirm the distance between the two wires and confirm that the pipe wall is uniform between the wires. Also confirm that there are no pipeline valves, fittings, or taps in between.

6.4.1.3 Calculate the resistance of the current span using Equations (6.7) to (6.9):

$$R = \rho \frac{L}{A}, \tag{6.7}$$

where

- R linear resistance of pipe for length L (ohms)
- ρ steel resistivity (ohm-centimeters)
- L length of current span (centimeters)
- A cross-sectional area of the pipe (centimeters squared)

or

$$A = \pi \frac{(\mathrm{OD}^2 - \mathrm{ID}^2)}{4}, \tag{6.8}$$

where

- OD outside diameter of the pipe (centimeters)
- ID inside diameter of the pipe (centimeters)

or

$$\mathrm{ID} = (\mathrm{OD} - 2wt), \tag{6.9}$$

where

- wt pipe wall thickness (centimeters)

6.4.1.4 Alternately, from Pipe Tables, determine the resistance of the pipe per unit length. The diameter and wall thickness or the weight per unit length must be known.

6.4.1.5 Multiply the length by the resistance per unit length to determine the overall resistance.

6.4.2 Four-Wire Span

6.4.2.1 From the results of the calibration tests, calculate the resistance of the pipe span from the test data using Equation (6.10):

$$R_p = \frac{[\mathrm{mV_{on}} - \mathrm{mV_{off}}]}{[\mathrm{I_{on}} - \mathrm{I_{off}}] \times 1{,}000} \tag{6.10}$$

where

R_p	resistance of pipe span (ohms)
$\mathrm{mV_{on}}$	voltage drop between inside test leads with current applied (millivolts)
$\mathrm{mV_{off}}$	voltage drop between inside pins with no current applied (millivolts)
I_{on}	current applied between outside test leads (amperes)
I_{off}	current interrupted, normally 0 A
1,000	converts millivolts to volts

6.4.2.2 Alternately, calculate a calibration factor ($\mathrm{CF_{span}}$) for the pipe span as follows:

$$\mathrm{CF_{span}} = \frac{[I_{on} - I_{off}]}{[\mathrm{mV_{on}} - \mathrm{mV_{off}}]} \tag{6.11}$$

where

$\mathrm{CF_{span}}$	calibration factor of pipe span (amperes per millivolt)
I_{on}	current applied between outside test leads (amperes)
I_{off}	current interrupted, normally 0 A
$\mathrm{mV_{on}}$	voltage drop between inside test leads with current applied (millivolts)
$\mathrm{mV_{off}}$	voltage drop between inside pins with no current applied (millivolts)

6.4.3 Pipeline Current

6.4.3.1 Using the current span resistance, calculate the pipeline current at each location using Equation (6.12):

$$I_{\text{span}} = \frac{V_{\text{span}}}{R_{\text{span}}}, \tag{6.12}$$

where

I_{span}	current in the pipeline current span (amperes)
V_{span}	voltage drop across the current span (volts)
R_{span}	resistance of the current span (ohms)

6.4.3.2 Alternately, using the current span calibration factor, calculate the pipeline current at each location using Equation (6.13):

$$I_{\text{span}} = \text{mV}_{\text{span}} \times \text{CF}_{\text{span}}, \tag{6.13}$$

where

I_{span}	current in the pipeline current span (amperes)
mV_{span}	voltage drop across the current span (millivolts)
CF_{span}	calibration factor of the current span (amperes per millivolt)

6.4.3.3 The current pickup (discharge) for any given section of pipeline between locations where current and direction are measured is given in Equation (6.14):

$$I_{\text{section}} = I_1 - I_2, \tag{6.14}$$

where

I_{section}	current pickup (discharge) from pipe section (amperes)
I_1	current measured at location 1 (amperes)
I_2	current measured at location 2 (amperes)

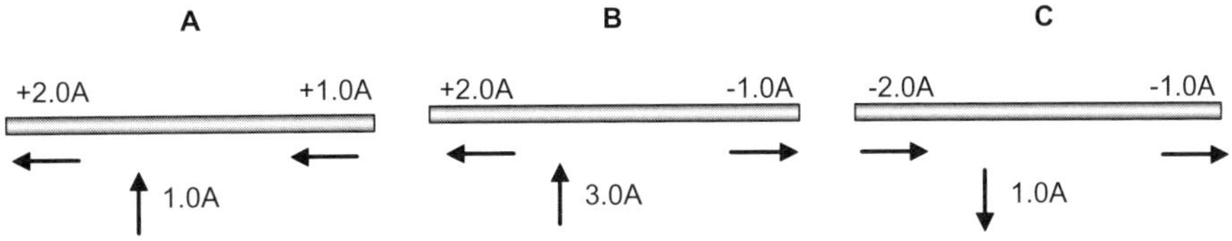

Figure 6.1 Examples of pipe current direction affecting current pickup or discharge are shown.

A common system of polarity (current direction) is necessary to determine the actual current pickup or discharge for each pipe section. Figure 6.1 shows the same current measured at each end, but the polarity or direction of current is different in each case, and the net current to or from the current is, therefore, different.

Using Equation (6.14) for each example in Figure 6.1,

$$\begin{aligned} I_{\text{section A}} &= +2.0\text{ A} - 1.0\text{ A} \\ &= +1.0\text{ A} \\ &\text{(pickup)}, \end{aligned}$$

$$\begin{aligned} I_{\text{section B}} &= +2.0\text{ A} - (-1.0\text{ A}) \\ &= +3.0\text{ A} \\ &\text{(pickup)}, \end{aligned}$$

$$\begin{aligned} I_{\text{section C}} &= -2.0\text{ A} - (-1.0\text{ A}) \\ &= -1.0\text{ A} \\ &\text{(discharge)}. \end{aligned}$$

7.0 REFERENCES

7.1 A.W. Peabody, *Control of Pipeline Corrosion*, 2nd ed., ed. R.L. Bianchetti (Houston, TX: NACE, 2001), p. 77.

7.2 M.E. Parker, *Pipeline Corrosion and Cathodic Protection*, ed. E.G. Peattie, 3rd ed. (Houston, TX: Gulf Publishing Company, Book Division, Houston, London, Paris, Tokyo, 1984), p. 31.

CATHODIC PROTECTION PROCEDURE NO. 4

Diagnostic Testing (Current Requirement) 4

1.0 INTRODUCTION

The purpose of diagnostic testing (DT) is to outline a process to test for deficiencies in a cathodic protection (CP) system, to identify the cause(s) of such deficiencies, and to determine the requirements to enhance the level of CP to the structure.

In an existing CP system, the DT is to first determine if the cause of low CP levels can be eliminated and, if not, what additional current is required, in addition to how and where it can be applied.

The CP Technician must have a good understanding of the criteria as well as the special conditions and precautionary notes that apply to these criteria.[1]

2.0 TOOLS AND EQUIPMENT

The following equipment will vary, depending on the test selected:

- Multimeter capable of measuring 1 mV_{DC} to 40 V_{DC}, complete with leads with insulated probes
- Copper-copper sulfate reference electrode
- Isolation checker
- Soil resistivity meter, complete with wires and four pins

[1] *See* Cathodic Protection Procedure No. 2: Structure-to-Electrolyte Potential Measurement.

- Multimeter, including alternating current (AC)/direct current (DC) volts and an ohmmeter
- Current interrupter
- DC ammeter sized for test current
- Battery and control resistor or portable controlled DC power supply
- Pipe locator transmitter and receiver
- Test wires as necessary for the applicable test
- Small hand tools

3.0 SAFETY EQUIPMENT

- Standard safety equipment and clothing, as required by the company's safety manual and regulations
- Electrically insulated clips and probe handles for meter leads
- Only personnel who have received training and are qualified in accordance with local codes and regulations are to work on DC power sources or their supply

4.0 PRECAUTIONS

The following precautions are in addition to those that must also be followed when working on a particular facility.

4.1 Measure the voltage between the rectifier case and ground before touching the case.

4.2 Open the case expecting to find biting insects, rodents, or snakes inside and take the appropriate precautions.

4.3 Inspect the rectifier for abnormal sounds, temperature, or odors and, if noted, turn it off.

4.4 Switch off the AC voltage supply before installing a current interrupter or each time the taps are adjusted.

4.5 Secure any exposed electrical terminals in a locked container when the rectifier is not attended.

4.6 Measure a structure AC voltage to ground on the structure before taking CP measurements. If the AC voltage to ground is equal to or exceeds 15 V_{AC}, safety measures detailed in NACE SP-0177[7.1] must be followed, and other personnel working on the structure of the hazard must be advised of such.

4.7 When working near high-voltage AC (HVAC) power lines, take AC structure-to-ground voltage readings at frequent intervals as these voltages can change with the power line load and geometry.

4.8 Do not work on the structure when lightning is in the area.

4.9 When working near a fence, confirm that it is not an electric fence for livestock (look for insulators) and that an AC voltage is not being induced on it by a parallel HVAC power line.

5.0 PROCEDURE

5.1 Information Required Prior to Diagnostic Testing

5.1.1 Previous annual survey data
5.1.2 Rectifier routine monitoring and annual data
5.1.3 Drawings
 5.1.3.1 Structure details
 5.1.3.2 CP installation details and location
 5.1.3.3 Test station types and locations
 5.1.3.4 Bond details and locations
5.1.4 Hazardous AC voltage tests (if applicable)
5.1.5 DC interference test results and mitigation effectiveness (if applicable)
5.1.6 Critical bond monitoring data
5.1.7 Isolation information
5.1.8 Road and railroad casing data (if applicable)
5.1.9 Close interval potential survey data (if applicable)
5.1.10 Pipeline coating information
5.1.11 In-line inspection or other inspection results (if applicable)
5.1.12 DT testing results

5.2 Investigate the Cause of Poor Cathodic Protection

5.2.1 Inspect the DC power sources and compare with the target DC volts and DC amperes. If the target outputs are significantly different from the target, complete the following inspections:

5.2.1.1 If there is 0 to ~2 V_{DC}[(2)] and 0 A_{DC} output, look for trouble in the rectifier or the AC supply to the rectifier. Confirm that the ~2 V_{DC} reading is from the DC power source by turning it off and disconnecting one DC cable. If the reading stays at ~2 V_{DC}, then it is from the DC power source. If it drops to 0 V, then it is the galvanic difference between the structure material and the anode or the carbon in the coke breeze.

5.2.1.2 If there is a normal voltage output but 0 A_{DC} output, look for the trouble in the cables, anodes, structure, or connections external to the rectifier.

5.2.1.3 If there is approximately one-half normal voltage and approximately one-half normal current, then investigate the possibility of a failed diode, causing the rectifier to half-wave.

5.2.1.4 If a problem exists in the rectifier or the external DC circuit,[(3)] complete repairs before proceeding with DT.

5.2.2 Inspect the DC bonds and repair any broken bonds found.

5.2.3 Test all isolating features.[(4)]

5.2.4 Test all road or railroad casings, if applicable, to confirm that they are isolated.[(5)]

5.2.5 If the problems causing the loss of CP, when corrected, restore protection, then the DT is completed.

5.2.6 When the preceding faults are corrected but a CP criterion has still not been met, proceed with the DT.

5.3 Structure-to-Electrolyte Potentials

5.3.1 Use structure-to-electrolyte potentials[(6)] to determine if the criterion for CP is being met.[7.2,7.3]

---

[2] Approximately 2 V_{DC} may be the galvanic difference between the steel in the structure and the carbon in the coke breeze, indicating that the DC power source is off.

[3] *See* Cathodic Protection Procedure No. 1: Rectifier Adjustment, Inspection, and Basic Troubleshooting.

[4] *See* Cathodic Protection Procedure No. 9: Electrical Isolation.

[5] *See* Cathodic Protection Procedure No. 10: Road and Railroad Cased Crossings (Basic).

[6] *See* Cathodic Protection Procedure No. 2: Structure-to-Electrolyte Potential Measurement.

5.3.2 Take structure-to-electrolyte potentials with a high-input impedance voltmeter (10 MΩ minimum) in conjunction with a copper-copper sulfate ($Cu/CuSO_4$) reference electrode (CSE).

5.3.3 Calibrate the field CSE by measuring a potential to a new, clean standard CSE that was recently charged with distilled water and copper sulfate crystals.[7] To calibrate the CSE, place the field CSE and the standard CSE reference electrodes in a nonmetallic container of clean water (alternately, hold the two porous plugs together) and measure the DC potential difference between them, using a good-quality voltmeter set on the low voltage scale. Service the field reference electrode, or replace it if the potential difference is greater than 5 mV.

5.3.4 Connecting the voltmeter positive to the structure and the negative to the reference electrode is now the preferred method. When connected in this manner, structure-to-electrolyte potential readings should then be negative and recorded as such. Connecting the voltmeter with the voltmeter negative to the structure is still permissible; however, the tester must realize that when a positive structure-to-electrolyte potential value is obtained, the reading is negative and must be recorded as such.

5.3.5 In rocky, sandy, very dry soils, or frozen ground, add water to the ground surface or a damp sponge attached to the reference electrode. In extreme conditions, a multiple-input impedance interface or multi-input impedance meter may be used. The potential measurements at a minimum of two input impedances must be the same; otherwise, the reference cell circuit resistance must be further reduced.

5.3.6 Document all techniques and raw data used to improve the quality of the data.

5.3.7 Take instant OFF structure-to-electrolyte potentials, with all influencing DC power sources interrupted, wherever practical.

5.3.8 Record the instant OFF potential reading between 0.6 and 1.0 s after interruption. If using a digital voltmeter, record the second

---

7 *See* Cathodic Protection Procedure No. 2: Structure-to-Electrolyte Potential Measurement, Appendix A.

reading displayed after interruption as the first display may be an average of the dropping values from the ON potential.

5.3.9 Except for close interval potentials, the spacing of the test station structure-to-electrolyte (pipe-to-soil) potential measurements should be from 2 to 3 km.

5.4 Dynamic Stray Current

5.4.1 Confirm that the structure under testing is not affected by a dynamic DC stray current.[(8)]

5.4.2 Retest adjacent potential readings that change in excess of 20% immediately to confirm proper reference electrode contact with the ground by the addition of water or by exposing moist soil.

5.4.3 Telluric or other dynamic stray current activity that requires calibration can be defined as an OFF potential fluctuation exceeding 20 mV peak to peak over the duration of the testing.

5.4.4 If a telluric or another dynamic stray current is detected, install data loggers at each end of the section to be tested to record structure-to-electrolyte (pipe-to-soil) potentials versus time and leave them recording for a period of approximately 22–24 hours, where practical.

5.4.5 Alternately, measure structure-to-electrolyte potentials manually and record the values and the time of each reading. Plot the results to see any trends.

5.4.6 If the test section is less than 1.6 km (1.0 mi), a single data logger may be installed at the test site.

5.4.7 Record each test station structure-to-electrolyte potential with another data logger for a period of 5 minutes (min). Take CIS potentials as normal, except the readings are to be time stamped.

5.4.8 Figure 5.1 illustrates that the wave prints of portable and stationary data loggers 10.5 km (6.5 mi) apart can be very similar. Although there was a small amount of telluric current in this time period, it is also demonstrated that a current interrupter failed in

[8] *See* Cathodic Protection Procedure No. 2: Structure-to-Electrolyte Potential Measurement, Sections 5.7 and Section 6.2; and *see* Cathodic Protection Procedure No. 8: Direct Current Stray Current Interference.

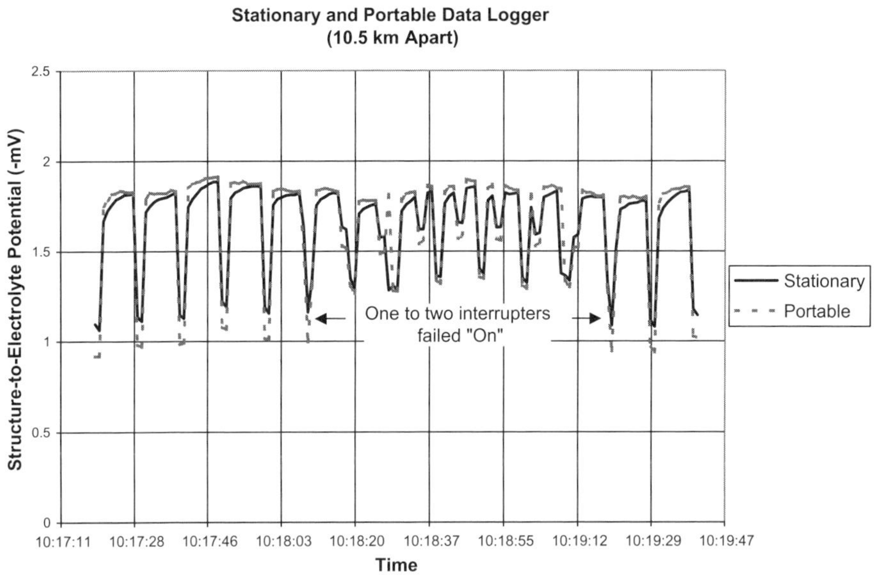

Figure 5.1 Comparison of wave prints of two data loggers 10.5 km (6.5 mi) apart, indicating current interrupter failure in addition to some telluric activity, is shown.

the ON position, typical of some Global Positioning System (GPS) interrupters, and another went out of synchronization; therefore, an incorrect reading may have been taken manually in this time frame without this knowledge.

5.5 Direct Current Power Source and Interrupter Installation

5.5.1 Record the nameplate data, tap setting, voltage, and current output of existing impressed current DC power sources and circuits in both the as-found and test conditions.

5.5.2 For sacrificial anode systems, record the size of the shunt, the millivolt reading measurement across it, and the current output.

5.5.3 Turn the DC power supply off and install a current interrupter in either the AC supply, the AC secondary taps, or the DC output of all influencing rectifiers, as shown in Figure 5.2, or in the DC output of other DC power sources. Install an interrupter in series with the sacrificial anodes and any bonds.

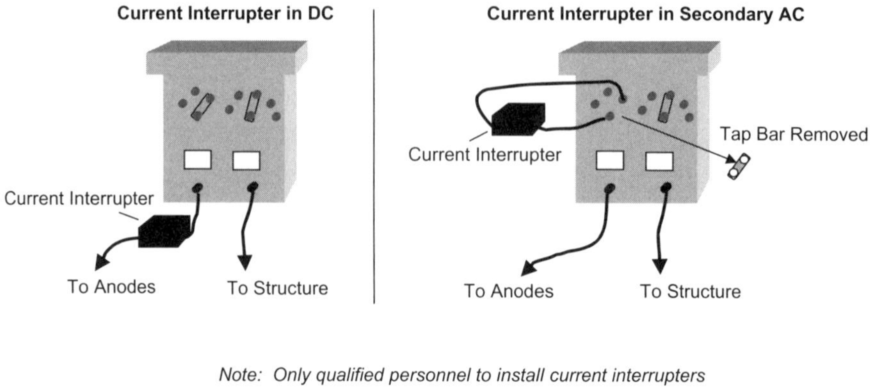

Figure 5.2 Typical current interrupter installations are shown.

5.5.4 For more than one current source, use synchronized interrupters, preferably GPS time-synchronized interrupters.
5.5.5 Select a long ON and a short OFF cycle to minimize the loss of polarization during the period of interruption, and record the timing of the cycles.
5.5.6 Adjust the rectifier, as necessary.[9]
5.5.7 If the feature exists, program the interrupters to turn off after the survey day ends and start again just before it begins to further maintain polarization.

5.6 Close Interval Structure-to-Electrolyte Potential

5.6.1 A detail procedure is given in "Cathodic Protection Procedure No. 7: Close Interval Potential Survey."
5.6.2 Connect the structure to the wire reel and move the reel with the voltmeter and reference electrode, as shown in Figure 5.3. In this way, a small break in the wire will still be part of the protected structure. A break in the insulation to the reference electrode will introduce an error in the reading when it comes in contact with the electrolyte.

[9] *See* Cathodic Protection Procedure No. 1: Rectifier Adjustment, Inspection, and Basic Troubleshooting.

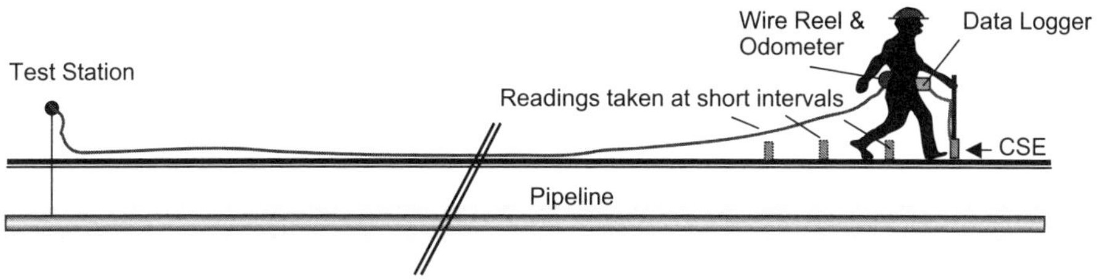

Figure 5.3 Close interval structure-to-electrolyte potential survey with trailing wire connected to structure, as recommended, is shown.

5.6.3 Take a structure-to-electrolyte potential directly to the structure and compare it to one taken to the wire in the reel, also connected to the structure. A difference in the reading will indicate a poor connection.

5.6.4 Keep the spacing of the reference cell 10 m (30 ft) or less in remote areas and no greater than 3 m (10 ft) in urban areas.

5.6.5 The direction of each survey segment should be recorded in the data.

5.6.6 The trailing wire is to be reconnected to the next test station or pipeline appurtenance, but prior to abandoning the previous pipe connection, a pipe-to-electrolyte potential measurement is to be taken to both the trailing wire and the next test station, with the reference cell in the same position. Both measurements are to be recorded. A small difference in these readings is to be expected due to the *IR* drop in the pipeline to the prior test station. If the new pipe connection is determined to be faulty, then the close interval survey may be continued using the previous pipe connection and noting this fact in the data. The voltmeter connection must not be made to a current carrying wire.

5.7 Baseline Survey

5.7.1 Obtain a complete set of baseline data, which may include but is not limited to the following:

5.7.1.1 AC structure-to-ground voltages measured prior to the DC structure-to-electrolyte potentials

5.7.1.2 ON/OFF structure-to-electrolyte potential tests, with all influencing current sources being interrupted to relate to the polarized potential criterion. All current sources include DC power sources (rectifiers; thermoelectric generators; and solar, wind, and engine generators), sacrificial anodes, and bonds. A close interval structure-to-electrolyte potential survey (CIS)[10] is advisable to confirm that all low-potential areas are addressed.

5.7.1.3 ON structure-to-electrolyte potential tests where current sources cannot be interrupted. In this case, additional testing must be completed to predict the *IR* drop error.[11] A CIS survey is advisable.

5.7.1.4 Depolarization potential tests (preferably a CIS), in which the current is left off after an ON/OFF potential test, are advisable.

5.7.1.5 DC current source outputs and bond current data

5.7.1.6 In addition to the CIS on the structure, measurement of structure-to-electrolyte potentials on each side of isolating features, on foreign structures, and on road or railroad casings

5.7.1.7 Comparison of the baseline survey data to the latest survey results to confirm that the location being tested is the intended area and that the operation of the CP system is similar

5.7.2 ON/OFF Structure-to-Electrolyte Potential Test

5.7.2.1 Determine the time cycle of the ON and OFF structure-to-electrolyte potential measurements.

5.7.2.2 Interrupt all influencing DC current sources on a recorded, timed ON and OFF cycle, preferably a long ON cycle to minimize the loss of polarization.

5.7.2.3 Install a stationary data logger to witness the interruption cycles and to confirm that all interrupters continue to operate and remain in synchronization. The stationary recorder will also serve to verify synchronized

[10] *See* Cathodic Protection Procedure No. 7: Close Interval Potential Survey.

[11] *See* Cathodic Protection Procedure No. 2: Structure-to-Electrolyte Potential Measurement.

interruption and if depolarization occurred during the survey interruption period.

5.7.2.4 Measure ON/OFF structure-to-electrolyte potentials and identify each location with a stake or paint so that the reference electrode can be placed in the exact position for subsequent tests.

5.7.3 ON Structure-to-Electrolyte Potentials

5.7.3.1 Determine the frequency of structure-to-electrolyte potential measurements.

5.7.3.2 Measure ON structure-to-electrolyte potentials and identify each location with a stake or paint so that the reference electrode can be placed in the exact position for subsequent tests.

5.7.3.3 Determine the *IR* drop component that is included in each ON structure-to-electrolyte potential.[(12)]

5.7.3.4 Calculate the true polarized potential by removing the *IR* drop error from the ON potential that was measured.

5.7.4 Depolarization Potentials

5.7.4.1 Determine the frequency of structure-to-electrolyte potential measurements.

5.7.4.2 Interrupt all influencing DC current sources on a long ON and OFF cycle and record the time for each cycle.

5.7.4.3 Measure ON/OFF structure-to-electrolyte potentials and identify each location with a stake or paint so that the reference electrode can be placed in the exact position for subsequent tests.

5.7.4.4 Turn off all current sources and record spot structure-to-electrolyte potentials over time, until the potentials have become relatively stable. In some cases, this may take up to several days. A stationary data logger will facilitate this test.

5.7.4.5 When the potentials have stabilized or reached the desired values, measure the depolarized potential with the reference electrode placed in the same exact locations as

[12] *See* Cathodic Protection Procedure No. 2: Structure-to-Electrolyte Potential Measurement.

during the ON/OFF structure-to-electrolyte potential survey.

5.7.4.6 Calculate the depolarization at each measurement location using Equation (5.1):

$$\Delta E_{dp1} = E_{\text{off}\,1} - E_{\text{depol}}, \tag{5.1}$$

where

ΔE_{dp1} depolarization (millivolts)

$E_{\text{off}\,1}$ instant OFF structure-to-electrolyte potential (millivolts)

E_{depol} depolarized structure-to-electrolyte potential (millivolts)

5.7.5 During baseline and current drain tests, measure foreign structure-to-electrolyte potentials where they may be affected by the proposed new CP installations. Where foreign facilities are affected, make a sketch of the configuration of the foreign facilities relative to the protected structure.

5.7.6 With the foreign owner's permission, test other locations to note the effect of the proposed CP system. Make sketches of the foreign structure and reading locations.

5.7.7 Existence of bonds with foreign structures should be recorded, including current and its direction.

5.8 Auxiliary Current Drain Tests

5.8.1 Install a temporary anode bed, such as ground rods, at the location that is anticipated for a new current drain point.

5.8.2 Install an ammeter, a control resistor, a DC power source, and a current interrupter in series between the structure and a temporary anode bed (ground rods), with the positive terminal connected to the temporary anode bed, as shown in Figure 5.4. If an adjustable DC current source is used, then the control resistor is not required. The wires and connections are to be sized for the expected current.

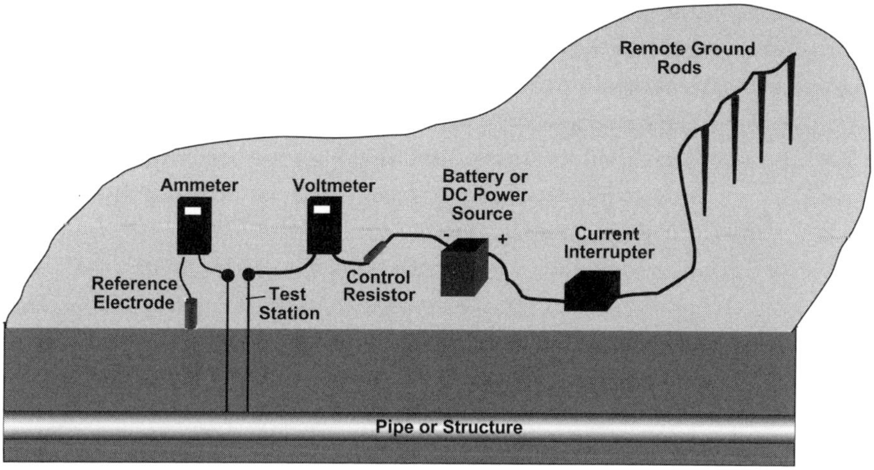

Figure 5.4 Typical temporary current drain test setup is shown.

5.8.3 Adjust the current to give a significant shift in the structure-to-electrolyte potential and confirm that the potentials are going more negative when the current comes on.

5.8.4 Take spot structure-to-electrolyte potentials at low CP points found in the baseline test to confirm that there is a response in the potentials from the temporary current being applied.

5.8.5 If the amount of polarization noted during this test does not exceed approximately 50 mV, then increase the current at the current drain test point and repeat the spot structure-to-electrolyte potentials. The amount of polarization will be the difference from the instant OFF potential less the baseline potential measured in the baseline survey (Equation [5.2]):

$$\Delta E_{p1} = E_{\text{off }1} - E_{\text{base}}, \tag{5.2}$$

where

ΔE_{p1}	polarization from DT temporary current
$E_{\text{off }1}$	instant OFF potential from DT temporary current
E_{base}	potential measured during baseline test

5.8.6 Measure ON/OFF structure-to-electrolyte potentials at each baseline reference electrode position, including isolating features, road casings, and foreign structures.

5.8.7 If a suitable response in structure-to-electrolyte potentials is not seen at all locations, either increase the current further or move the temporary current drain point to another location and repeat the preceding tests.

5.8.8 Calculate the true OFF structure-to-electrolyte potential by adding the amount of polarization in this test to the baseline polarized (instant OFF) structure-to-electrolyte potential.

5.9 Pipeline Current Measurements

5.9.1 Measure the current in a pipeline at intervals, if possible, to determine the distribution of current with the test current both ON and OFF.[13]

5.9.2 Pipeline Current Span

5.9.2.1 Calibrate the resistance or calibration factor of a four-wire pipeline current span, as described in "Cathodic Protection Procedure No. 3: Direct Current Measurements." Measure the millivolt drop across the two inside wires and calculate the current (Figure 5.5).

5.9.2.2 With a two-wire test station, determine the distance between the wires, the diameter, the wall thickness, and the material of the pipeline at that location to determine the resistance of the pipeline current span by resistance tables or by the material resistivity.

5.9.3 Clamp-On Ammeter

5.9.3.1 If the clamp-on ammeter is an AC/DC type, ensure that the setting is on DC.

5.9.3.2 Center the coil on the pipeline and record the ON/OFF current, the polarity of each, and, thus, the direction of the current (Figure 5.5).

---

13 *See* Cathodic Protection Procedure No. 3: Direct Current Measurements, Section 5.4.

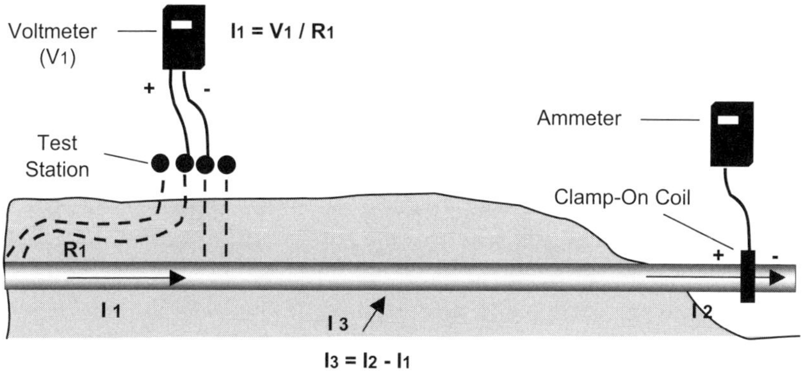

Figure 5.5 Two methods of measuring pipeline current are shown.

5.9.3.3 Record the current and the positive polarity as being upstream or downstream.

5.9.3.4 Reverse the coil and repeat the measurement to ensure that both readings are close and, therefore, representative.

5.9.3.5 Average the current readings for the final value, but record all data.

5.10 Coating Conductance[7.3]

5.10.1 Select a section where the current can be measured at each end, either by a current span or by a clamp-on ammeter.

5.10.2 Install a current interrupter in a nearby CP current source or install a temporary current drain, as shown in Figure 5.4.

5.10.3 An example of the tests to be completed is illustrated in Figure 5.6. Pipeline current can also be measured by a clamp-on ammeter, as shown in Figure 5.5.

5.10.4 Measure the ON/OFF current in the pipeline at one end (location 1 in Figure 5.6). If using a pipeline current span, measure the millivolt drop between the two inside wires of the span and calculate the current from Equation (5.3):

$$I = \frac{\text{mV}}{1{,}000R} \tag{5.3}$$

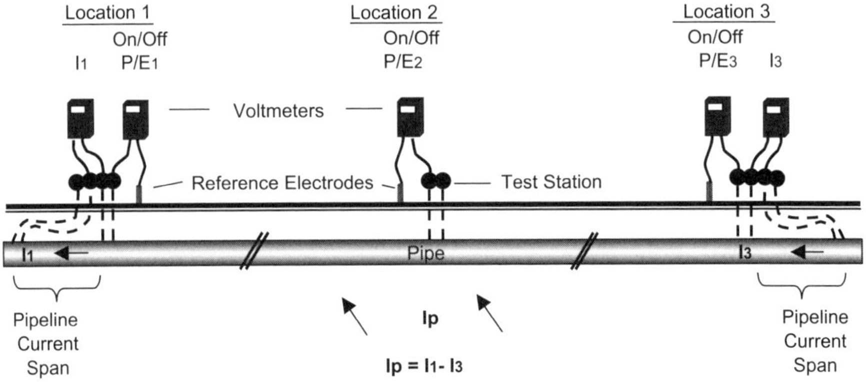

Figure 5.6 Coating conductance tests are shown.

where

I	current in a pipeline current span (amperes)
mV	voltage across pipeline current span
R	resistance of pipeline current span (ohms)
1,000	converts millivolts to volts

5.10.5 If using a clamp-on ammeter, take current readings in both directions to confirm that they are approximately the same and then calculate the average.

5.10.6 Measure the ON/OFF current in the pipeline at the other end (location 3 in Figure 5.6).

5.10.7 Note that there will likely be residual current in the pipe with the DC current source OFF. The current (I) of interest is the difference between the pipeline current with the test current ON (I_{on}) and with it interrupted (I_{off}) in each case (Equation [5.4]):

$$I = I_{on} - I_{off}. \tag{5.4}$$

5.10.8 Measure the ON/OFF structure-to-electrolyte potentials at each end and at accessible locations in between (locations 1, 2, and 3 in Figure 5.6).

5.10.9 If the current cannot be measured, or if the difference between the ON and OFF potentials is less than 50 mV, then apply more current and repeat the preceding tests.

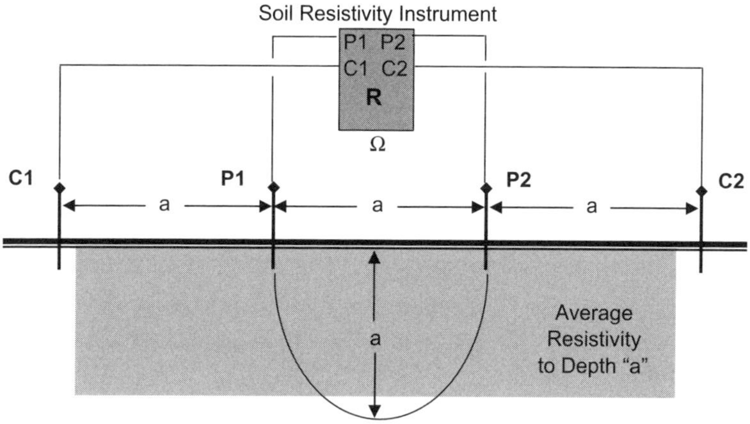

$$\rho = 2\pi a R$$

Figure 5.7 Soil resistivity using the four-pin method is shown.

5.11 Soil Resistivity

5.11.1 Take soil resistivity measurements(14) using the four-pin method, as shown in Figures 5.7 and 5.8, at each possible CP anode bed site. Make certain that the pins are connected exactly as shown.

5.11.2 If a remote anode bed is planned, obtain soil resistivity measurements in a 30-m (100-ft) grid pattern throughout the proposed anode site at each soil depth to a depth greater than the anticipated anode depth, and another to a depth between that and the surface. Extend the grid until the lowest uniform soil resistivity has been determined to encompass the area of the proposed anode bed (Figure 5.8).

5.11.3 If the proposed corrosion remedial design is a deep anode system, then take resistivity tests to different depths, with the deepest depth exceeding the anticipated total anode bed depth (e.g., 25 m, 50 m, 75 m, 100 m [82 ft, 164 ft, 246 ft, 328 ft]). Note that three times this distance on the surface is needed for the three equal spacings between the pins. The soil resistivity instrument may not be sensitive enough for deeper

14 *See* Cathodic Protection Procedure No. 12: Soil Resistivity Measurement.

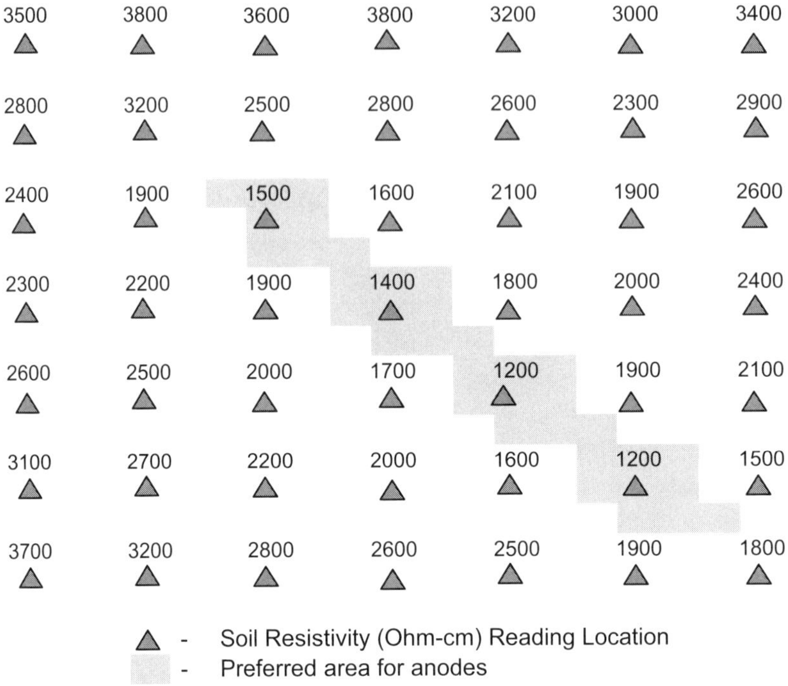

Figure 5.8 Soil resistivity grid pattern for anode location is shown.

measurements, in which case, refer to "Cathodic Protection Procedure No. 12, Soil Resistivity Measurement."

5.11.4 If a distributed anode system is planned, soil resistivity measurements are to be taken every 30 m (100 ft), and if there is a 2:1 or 1:2 change in resistivity between consecutive readings, reduce the spacing by half in those areas.

5.11.5 Soil resistivity pin alignment is to be perpendicular to a metallic structure when it is nearby.

5.12 Additional Cathodic Protection Design Information

5.12.1 Obtain the following information in addition to the preceding test data:

5.12.1.1 Drawing of the temporary current drain, showing location of temporary anode bed(s)

5.12.1.2 Drawing with dimensions of the proposed anode bed locations, with dimensions to structure, power lines, topographical features, foreign structures, roads, and landowner divisions

5.12.1.3 All underground or aboveground power lines and AC voltage (single phase or three phase) related to the structure

5.12.1.4 Structure or pipeline drawings showing the location of the possible anode beds, test stations, foreign structures (pipelines), power line locations, and access

5.12.1.5 Bodies of water (sloughs, creeks, rivers, lakes, swamps), especially at possible anode bed locations

5.12.1.6 Details of foreign structures and foreign CP systems

5.12.1.7 Description of soil and significant soil features (e.g., granite, muskeg)

5.12.1.8 Topography (hills, low areas)

5.12.1.9 The location and size of existing rectifiers and ground beds

5.12.1.10 Possible sources of AC or DC stray current

5.12.1.11 Photographs of the anode bed area and unusual features of the structure or area

6.0 ANALYSIS

6.1 Criteria to Be Met

6.1.1 CP criteria are detailed in Section 6 of NACE SP0169.[7.3] [(15)]

6.1.2 There are currently three structure-to-electrolyte potential criteria for submerged or buried steel structures in the absence of specific data that demonstrate that adequate CP has been applied, including the following:

6.1.2.1 A negative (cathodic) potential of at least 850 mV (with respect to a CSE contacting the electrolyte) with the cathodic protection applied but with voltage drops other than across the structure-to-electrolyte interface (*IR* drop)

---

15 Use most recent version of criteria.

removed. *IR* drops between the reference electrode and the structure-to-electrolyte boundary are an error in this reading and must be removed from the ON potential before applying this criterion, as illustrated by Equation (6.1). NACE SP0169[7.3] discusses methods to evaluate the *IR* drop, which is essentially the difference between the ON potential and the instant OFF potential:

$$E_c = (E_{\text{on}} - IR), \tag{6.1}$$

where

- E_c potential for criterion (millivolts; -850 mV_{CSE} or more negative)
- E_{on} potential with current applied (millivolts)
- IR voltage drop between the reference electrode and the structure-to-electrolyte boundary (millivolts)

6.1.2.2 A polarized potential equal to or more negative than 850 mV relative to a saturated copper-copper sulfate reference electrode contacting the electrolyte. This is obtained when all current sources influencing the structure have been temporarily interrupted and an instant OFF potential (E_{off}) is read.

6.1.2.3 A minimum of 100 mV of cathodic polarization between the structure surface and a stable reference electrode contacting the electrolyte. Polarization is the change in potential from the native or free corroding potential and the instant OFF potential, as illustrated by Equation (6.2). It can be measured during formation or decay after all influencing power sources have been turned off:

$$E_p = (E_{\text{off}} - E_{\text{native}}), \tag{6.2}$$

where

- E_p polarization for criterion (100 mV or greater)
- E_{off} potential with all current momentarily interrupted

E_{native} native potential before CP current is applied if in formation or depolarized potential after current is interrupted for a period of time if in decay.

6.1.3 Only one of these criteria needs to be met. For example, if a polarized potential is more electropositive than $-850\ mV_{CSE}$, a depolarization survey may prove that the 100-mV criterion is being achieved. See NACE SP0169[7.3] for special conditions.

6.2 Cause of Subcriterion Potentials

6.2.1 Some common causes of a system's inability to meet criteria, tests, and cures are listed in Table 6.1. Exceptions can always be encountered.

6.3 Structure-to-Electrolyte Potentials

6.3.1 Compare the structure-to-electrolyte potentials with the historical data.

6.3.2 Readings that are more electronegative than before and at the same current output may suggest that the structure has been reduced in size, possibly by a faulty bond, thus, isolating part of the structure, or that a CP system associated with another structure at a more electronegative potential is now shorted. The potential of the remaining structure will become more electropositive.

6.3.3 Readings that are more electronegative than before at the same current output may also suggest anodic interference; that is, a DC ground or anode bed of a foreign DC power source is in close proximity to the structure.

6.3.4 Readings that are less electronegative than before may indicate one or more of the following problems:

6.3.4.1 Low DC output from the DC power source due to the following:

6.3.4.1.1 Failing anodes

6.3.4.1.2 No AC power or energy source to the DC power supply

Table 6.1 Summary of Identification of Cathodic Protection Trouble, Tests, and Causes

System Component	DC Power Source		Secondary AC Voltage	Structure-to-Electrolyte Potentials	Trouble Suspected	Tests	Remedy
	Volts	Amperes					
Sacrificial Anode							
		A–		P–	May not be trouble. Anode current will decrease when potential becomes more negative.	No action	
		A+		P+	Opposite to above	Test for shorts or faults in the structure system.	Repair as required.
		A–		P+	Anodes failing	Measure anode-to-electrolyte potential. Perform anode voltage gradient test.	Replace anodes as necessary.
DC Power Source							
	0 to ~2*	0	0	P+	No AC power or failed DC power components. Trouble is before or in DC power source.	Confirm AC supply, check circuit breaker, test fuses, poor connections, or broken wires in DC power source. Check for signs of heat. If a battery supplemented source, test batteries for charge.	Confirm cause and correct before re-energizing.
						If circuit breaker trips, look for short.	If no short, reduce DC output voltage.

	0 to ~ 2*	0	0	P+	No secondary AC voltage at taps	Test AC supply and circuit breaker.	If AC supply and circuit breaker are OK, then test transformer.
	0 to ~ 2*	0	V	P+	Fuse(s) in DC power source, failed rectifying element, poor connections or faulty wire	Test fuse(s), rectifying element, connections, and wires.	If fuse, test for short. If none found, lower voltage and re-energize. Otherwise, replace or repair as required.
	~1/2 V	~1/2 A	V	P+	One-half wave DC output. One part of rectifier bridge circuit is open.	Turn off, remove rectifying element connections, and test each diode or element.	Replace rectifying element(s).
	?V	?A	V	P	Faulty meters	Calibrate meters.	Replace as necessary.
DC Cables/Anode Bed							
	V	0	V	P+	Faulty cable, connections, or anodes	Trace cable to structure and anodes. Perform anode voltage gradient test.	Repair or replace as required.
	V	Dropping over time	V	P+	Failing or dry anodes	Potential profile over anodes to confirm status. Turn off to note recovery.	Temporary cure may be to water anodes. Replace anodes as required.

(continued)

Table 6.1 *(continued)*

System Component	DC Power Source Volts	DC Power Source Amperes	Secondary AC Voltage	Structure-to-Electrolyte Potentials	Trouble Suspected	Tests	Remedy
Structure							
	V	A	V	P+	Shorted isolation, accidental contact to foreign structure, faulty bonds, deteriorating coating	Test isolation and bonds. Trace for contacts. Coating conductance test. Complete DT.	Repair isolation or bonds. Separate any contacts to foreign structures. Add CP capacity for poor coating or recoat.

Legend:

V Normal voltage

A Normal current

P Normal structure-to-electrolyte potential

V+ Greater than normal voltage

V− Lower than normal voltage

A+ Greater than normal current

A− Lower than normal current

?V, ?A Abnormal readings or varying

P+ Structure-to-electrolyte potential more electropositive

P− Structure-to-electrolyte potential more electronegative

* ~2 V may be due to galvanic difference between steel and anode or carbon in the coke breeze and not an indication of power

6.3.4.1.3 Faulty components in the DC power source

6.3.4.1.4 Faulty cables or connections in the external DC circuit

6.3.4.1.5 A short in the AC or DC circuitry

6.3.4.2 Shorted isolation, adding a foreign structure to the CP system and increasing the current requirements if the foreign system is at more electropositive potentials

6.3.4.3 Shorted casings, adding more bare metal to the structure and, therefore, increasing the current required for protection

6.3.4.4 A contact to a foreign structure, thereby increasing the current required as the structure-to-electrolyte potential on both structures has to be brought more negative simultaneously

6.3.4.5 Deteriorating coating, resulting in an increase in the current necessary to achieve criteria

6.3.4.6 A faulty interference bond or new interference from a foreign DC power source

6.4 Dynamic Stray Current

6.4.1 Telluric or other dynamic stray current activity requiring calibration is defined as an OFF potential fluctuation exceeding 20 mV peak to peak over the duration of the testing.(16)

6.4.2 Where data loggers have been installed, process the structure-to-electrolyte potential data collected and prepare a corrected profile. Note any repeating waveforms in the original profile suggesting a man-made source.

6.4.3 Match the portable data log potential profile to the stationary data log profile at the same precise time.

6.4.4 Determine the true potential at the stationary data loggers first, either by finding a quiet period in the data or, under moderate conditions, by taking an average over the test period. For each potential measurement along the line, calculate the difference in

[16] *See* Cathodic Protection Procedure No. 8: Direct Current Stray Current Interference.

potential between that and the stationary data logger at the same moment in time. This difference added to the difference to the true potential at the stationary data logger is the correction factor for the portable data logger reading. Equations are presented for the following situations:

- If two stationary data loggers are used, the method shown in Equations (6.3) and (6.4) can be used:

$$\acute{\varepsilon}_b = [\acute{\varepsilon}_a(c - b)/c] + [\acute{\varepsilon}_c(b - a)/c], \tag{6.3}$$

where

a	first stationary potential location
b	portable potential location
c	second stationary potential location
$\acute{\varepsilon}_a$	error in potential at a at time x
$\acute{\varepsilon}_b$	error in potential at b at time x
$\acute{\varepsilon}_c$	error in potential at c at time x

and

$$E_p = E_{p\ \text{measured}} - \acute{\varepsilon}_b, \tag{6.4}$$

where

E_p	true potential at the portable data logger location
$E_{p\ \text{measured}}$	potential at the portable data logger location

- If only one stationary data logger was used, then Equation (6.5) applies:

$$E_p = E_s - (E_{sa} - E_{pa}), \tag{6.5}$$

where

E_p	true potential at the portable data logger location
E_s	true potential at the stationary location
E_{sa}	stationary potential at time a during the data logging
E_{pa}	portable potential at time a during the data logging

6.4.5 Other methods can also be used to correct for dynamic stray currents.

6.4.6 The stationary recorder will also serve to verify synchronized interruption and if depolarization occurred during the survey interruption period.

6.5 Close Interval Potential Survey

6.5.1 Review the structure-to-electrolyte potentials[(17)] to determine if they meet one of the criteria[(18)] for CP throughout.[7.1]

6.5.2 In the event that the −850 mV_{CSE} is not met, review the depolarized potentials to determine if the 100-mV criterion is achieved. If a depolarized test was not conducted, determine the practicality of conducting this test.

6.5.3 The structure-to-electrolyte potentials will attenuate gradually away from the current drain point where the coating is uniform.

In Figure 6.1, more rapid attenuation is seen where there is a large amount of bare metal exposed to the soil electrolyte, either in the form of large coating holidays or another structure in contact. A recovery of the potentials will normally be seen when the better coating is reached. This voltage gradient is a reflection of the higher current density at the bare metal.

6.6 Direct Current Power Source Interruption

6.6.1 Review stationary data log profiles to confirm that the interrupters at all influencing rectifiers continued to function during the test (see Figure 5.1).

6.6.2 Identify any structure-to-electrolyte potential readings that may have been affected by an interrupter malfunction.

6.6.3 Note the amount of depolarization that took place during the interruption period.

---

[17] *See* Cathodic Protection Procedure No. 2: Structure-to-Electrolyte Potential Measurement.

[18] *See* Section 6.1.

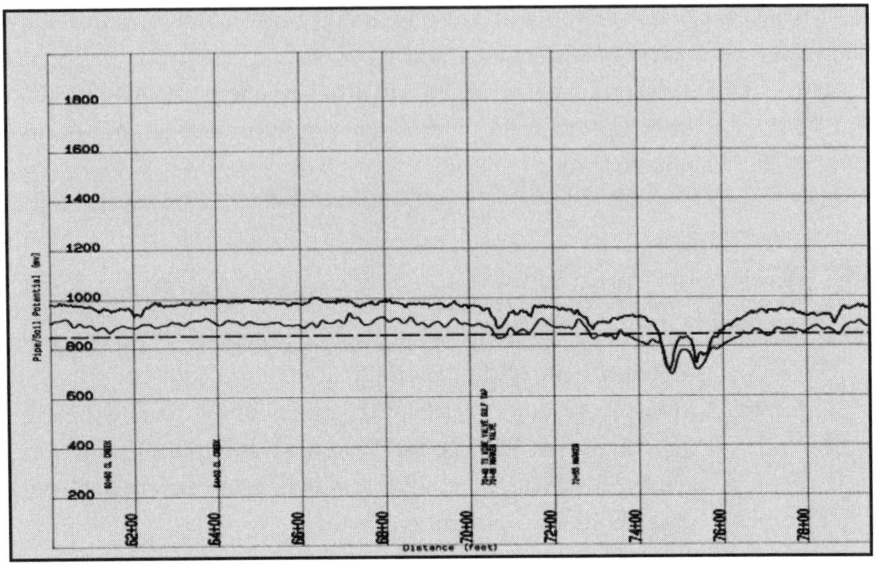

Figure 6.1 Typical close interval structure-to-electrolyte potential profile is shown. (*Source:* NACE CP Technician Course.)

6.7 Baseline Tests

6.7.1 Compare the results of the other DT tests with the initial baseline tests to note the effect of the repair work or the additional testing.

6.7.2 This baseline only applies to this DT survey, as the intent is to make improvements either by making repairs to the system or by adding to the capacity of the CP system.

6.7.3 A new baseline will be established after the DT recommendations have been implemented. Providing that the structure system is not modified, the current requirements have not increased, and the ON structure-to-electrolyte potential has not changed, the new baseline can be used for the following year or so.

6.8 Auxiliary Current Drain Tests

6.8.1 Where the temporary current drain test did not achieve the CP criteria but is relatively close, Equation (6.6) can approximate the current necessary to meet the -850-mV$_{CSE}$ criterion:

$$I_{\text{reqd}} = \frac{(-850\,\text{mV} - E_{\text{native}})}{(E_{\text{off}} - E_{\text{native}})} \times I_{\text{test}}, \tag{6.6}$$

where

I_{reqd}	current required to achieve –850-mV$_{CSE}$ criterion (amperes)
E_{native}	native structure-to-electrolyte potential (millivolts)
E_{off}	instant OFF (polarized) structure-to-electrolyte potential (millivolts)
I_{test}	test current applied for E_{off}(amperes)

Note: Include polarity of potential readings.

6.8.2 Equation (6.7) can approximate the current necessary to meet the 100-mV criterion from the temporary test current:

$$I_{\text{reqd}} = \frac{(-100\,\text{mV})}{(E_{\text{off}} - E_{\text{native}})} \times I_{\text{test}}, \tag{6.7}$$

where

I_{reqd}	current required to achieve the 100-mV criterion (amperes)
E_{off}	instant OFF (polarized) structure-to-electrolyte potential (millivolts)
E_{native}	native structure-to-electrolyte potential (millivolts)
I_{test}	test current applied for E_{off} (amperes)

Note: Include polarity of potential readings.

6.9 Pipeline Current Measurement

6.9.1 Current Span Method (see also Cathodic Protection Procedure No. 3 Direct Current Measurements, Section 6.4).

6.9.1.1 To facilitate this type of test, special current span test leads must exist. A current span must be located at each end of the pipe section being tested.

6.9.1.2 A four-wire current span is to be calibrated first by measuring the resistance of the span in ohms or

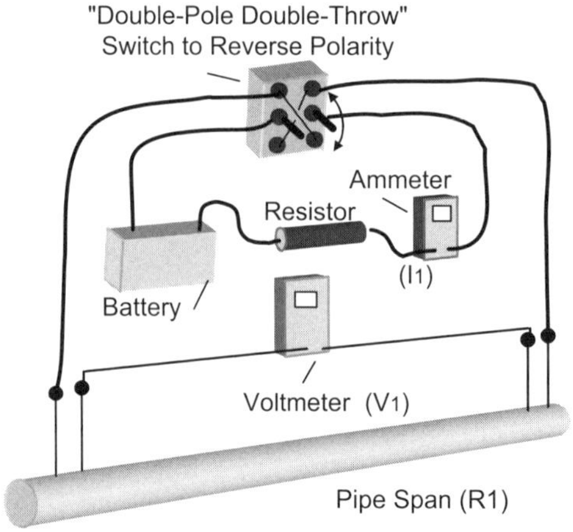

Figure 6.2 Pipe current span is shown.

calculating a calibration factor for the span in amperes per millivolt.

6.9.1.3 Calculate the resistance of the pipe span (Figure 6.2) by Ohm's law from Equation (6.8):

$$R_1 = \frac{\Delta V_1}{\Delta I_1}, \tag{6.8}$$

where

R_1 resistance of the pipe span (ohms)

ΔV_1 net test voltage across the pipe span ($V_{on} - V_{off}$; volts)

ΔI_1 net test current through the span ($I_{on} - I_{off}$; amperes)

6.9.1.4 Alternately, calculate the calibration factor from Equation (6.9):

$$\mathrm{CF}_1 = \frac{\Delta I_1}{\Delta \mathrm{mV}_1}, \tag{6.9}$$

where

CF_1 calibration factor of span 1 (amperes per millivolt)

ΔI_1 net current through the span (amperes)

$\Delta \mathrm{m} V_1$ net voltage across the pipe span (millivolts)

6.9.1.5 Where two-wire spans exist, calculate the resistance from the pipe dimensions and the resistivity of steel ($\sim 18 \times 10^{-6}$ Ωcm), as shown in Equations (6.10), (6.11), and (6.12). These equations are shown in metric units; however, imperial units can be used, provided they are all consistent in type of unit:

$$R_{\text{span}} = \rho \frac{L_{\text{span}}}{A}, \tag{6.10}$$

$$A = \pi \frac{(\text{OD}^2 - \text{ID}^2)}{4}, \tag{6.11}$$

$$\text{ID}_{\text{pipe}} = \text{OD}_{\text{pipe}} - 2wt, \tag{6.12}$$

where

R_{span}	steel pipe resistance (ohms)
ρ	resistivity of steel (ohm-centimeters)
A	cross-sectional area of pipe (centimeters squared)
wt	wall thickness (centimeters)
OD_{pipe}	pipe outside diameter (centimeters)
ID_{pipe}	pipe inside diameter (centimeters)
L_{span}	length of span (centimeters)
π	constant 3.1416

Note: 1 in = 2.54 cm = 25.4 mm; 1 ft = 30.48 cm = 0.3048 m; 1 m = 100 cm = 39.37 in = 3.27 ft.

6.9.1.6 Alternately, calculate a calibration factor *CF* in amperes per millivolt, as shown by Equation (6.13):

$$F_1 = \frac{\Delta I_1}{\Delta \text{mV}_1}, \tag{6.13}$$

where

F_1	calibration factor of the pipe span (amperes per millivolt)
ΔI_1	net current through the span (amperes)
ΔmV_1	net voltage across the pipe span (millivolts)

6.9.1.7 Current is calculated for a pipeline current span either from Equation (6.14), if resistance is known, or from Equation (6.15), if the calibration factor is known. Note the units in each case.

- Using resistance,

$$I_1 = \frac{V_1}{R_1}, \tag{6.14}$$

where

I_1 current through the span (amperes)

V_1 voltage across the pipe span (volts)

R_1 resistance of the pipe span (ohms)

- Using the calibration factor,

$$I_1 = \mathrm{mV}_1 \times F_1, \tag{6.15}$$

where

I_1 current through the span (amperes)

mV_1 voltage across the pipe span (millivolts)

F_1 calibration factor of the pipe span (amperes per millivolt)

Note: The voltage must be in volts if using the resistance but in millivolts when using the calibration factor.

6.9.1.8 Record the current direction.

6.9.2 Clamp-On Ammeter

6.9.2.1 Average the current taken with the coil in both directions.

6.9.2.2 Record the current direction.

6.10 Coating Conductance[7.3]

6.10.1 Equation (6.16) gives the coating conductance relationship to coating resistance:

$$G = \frac{1}{R}, \tag{6.16}$$

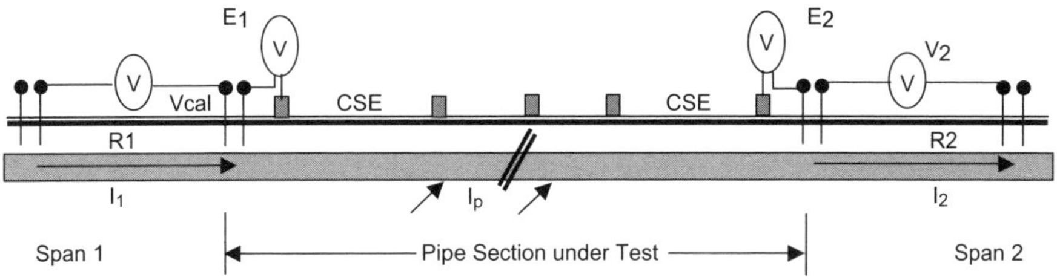

Figure 6.3 Coating conductance test is shown.

where

G conductance (siemans)

R resistance (ohms)

6.10.2 Note that conductance is the inverse of resistance. Either calculating the resistance and taking the inverse or calculating directly from the data can determine the coating conductance. The latter approach is described subsequently.

6.10.3 Determine the current applied only to the section in question.

6.10.4 Calculate the average potential change (ΔE) through the pipe section as a result of the applied current (Figure 6.3).

6.10.5 The current applied is then calculated as follows:

- Current for an isolated section is equal to the current applied.
- Current pickup to the pipeline coating conductance section (see Figure 6.3) is determined by Equation (6.17) from the current at each end of the section:

$$\Delta I_p = [(I_{\text{on}\,1} - I_{\text{off}\,1}) - (I_{\text{on}\,2} - I_{\text{off}\,2})], \tag{6.17}$$

where

ΔI_p net current to the section (amperes)

$I_{\text{on}\,1}$ current at span 1 with test current on (amperes)

$I_{\text{off}\,1}$ current at span 1 with test current off (amperes)

$I_{\text{on}\,2}$ current at span 2 with test current on (amperes)

$I_{\text{off}\,2}$ current at span 2 with test current off (amperes)

6.10.6 The coating conductance is normally calculated as specific coating conductance, that is, the conductance of a specific unit area of the pipe surface. The specific coating conductance in siemans per meter squared of a given pipe section is calculated by Equation (6.18):

$$g' = \frac{\Delta I_p}{\Delta E \times A} \tag{6.18}$$

where

- g' specific coating conductance of pipe section (siemans per meter squared)
- ΔI_p net current to pipe section (amperes)
- ΔE average difference in ON and OFF pipe-to-electrolyte potentials throughout the pipe section (volts)
- A surface area of pipe section picking up current (meters squared); $A = \pi dL$, where π is the constant 3.141593, d is diameter (meters), and L is length (meters)

6.11 Soil Resistivity

6.11.1 Calculate the soil resistivity[(19)] from the four-pin test using Equation (6.19):

$$\rho = 2\pi aR, \tag{6.19}$$

where

- ρ resistivity (ohm-centimeters)
- π constant 3.141593
- a inside pin spacing and average depth of measurement (centimeters)
- R resistance measured or calculated (ohms)

6.11.2 Calculate the resistivity from a soil (water) box using Equation (6.20):

$$\rho = \frac{RA}{L}, \tag{6.20}$$

[19] Four-pin calculations described in more detail in "Cathodic Protection Procedure No. 12: Soil Resistivity Measurement."

where

- ρ resistivity (ohm-centimeters)
- A cross-sectional area of soil box (centimeters squared)
- L distance between potential pins, that is, P1 and P2 (centimeters)
- R resistance measured or calculated (ohms)

Note: If the soil box is constructed where $A = L$, then Equation (6.20) becomes Equation (6.21):

$$\rho = R, \tag{6.21}$$

where

- ρ resistivity (ohm-centimeters)
- R resistance measured or calculated (ohms)

Although in Equation (6.21), the value of the resistivity is equal to the resistance, the resistivity is in ohm-centimeters, and the resistance is in ohms.

6.11.3 Soil resistivity layers can be predicted by the Barnes layer method.(20) Figure 6.4 illustrates three layers of soil resistivity that are to be predicted based on the average soil resistivity measured to three different depths (a_1, a_2, and a_3).

The Barnes layer analysis(21) assumes that the layers are uniform and in parallel. A_{Layer} is the thickness of each layer, which is the difference between the two depths measured. The resistance R_{Layer} of each layer can be calculated by the parallel resistance formula described for each layer in Figure 6.4. With the spacing and the resistance known, the resistivity (ρ_{Layer}) can be calculated from Equation (6.19) or as shown in Figure 6.4.

[20] The Barnes layer method is described in more detail in "Cathodic Protection Procedure No. 12: Soil Resistivity Measurement."

[21] *See* Cathodic Protection Procedure No. 12: Soil Resistivity Measurement.

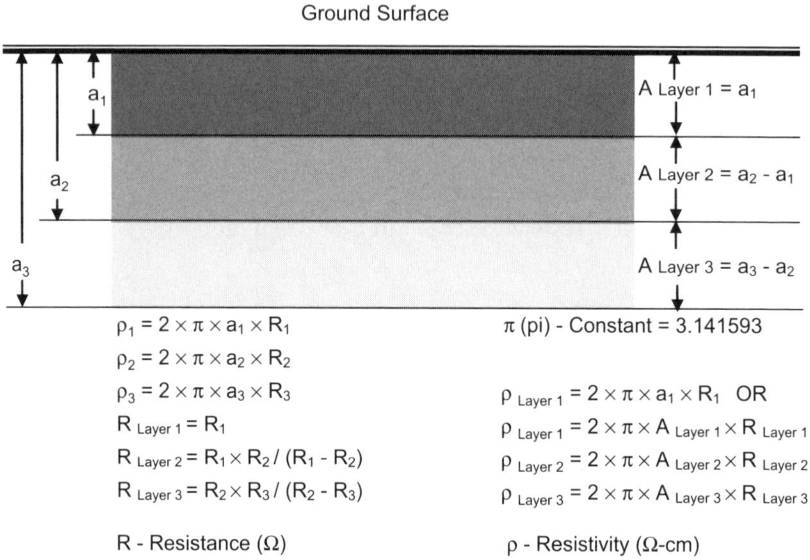

Figure 6.4 Soil resistivity layers with calculations are shown.

6.12 Additional Cathodic Protection Design Information

6.12.1 This section is only intended to explain how the field information will be used in a CP design, rather than to describe how to prepare a CP design.

6.12.2 The configuration of the temporary anode bed will be compared to that anticipated in the new anode bed design to decide how the distribution of current will change.

6.12.3 Details of the proposed anode bed locations are required to prepare the design and a list of materials and to acquire land. Anodes are to be placed in the following:

- 6.12.3.1 The lowest soil resistivity
- 6.12.3.2 Uniform soil resistivity
- 6.12.3.3 Permanent moisture
- 6.12.3.4 A layer below the maximum frost depth

6.12.4 Normally, the most suitable locations for anode beds are in geographically low areas, providing the soil resistivity is suitable.

6.12.5 In addition to deciding if an AC voltage supply is practical, the possibility of future hazardous AC voltages has to be considered.

6.12.6 Anode beds are to be installed away from foreign structures to avoid interference effects on them. Also, details of the foreign CP systems are needed to make certain that they will not interfere either.

6.12.7 Any sources of AC or DC stray current need to be addressed as they will not only affect the design but also the commissioning survey.

6.12.8 Photographs of the proposed anode bed area(s) and unusual features of the structure or terrain can prove invaluable in the preparation of a design.

7.0 REFERENCES

7.1 NACE Standard SP0177-2007, "Mitigation of Alternating Current and Lightning Effects on Metallic Structures and Corrosion Control Systems" (Houston, TX: NACE International, 2007).

7.2 NACE Standard TM0497-2002, "Measurement Techniques Related to Criteria for Cathodic Protection on Underground or Submerged Metallic Piping Systems" (Houston, TX: NACE International, 2002).

7.3 NACE SP0169-2007, "Control of External Corrosion on Underground or Submerged Metallic Piping Systems" (Houston, TX: NACE International, 2007).

7.4 NACE Standard TM0102-2002, "Measurement of Protective Coating Electrical Conductance on Underground Pipelines" (Houston, TX: NACE International, 2002).

CATHODIC PROTECTION PROCEDURE NO. 5

Adjustive Survey 5

1.0 PURPOSE

The purpose of the adjustive survey is to confirm that the cathodic protection (CP) system continues to operate meeting its objective, or if not, to adjust it to achieve a CP criterion. If this cannot be achieved within the capacity of the CP system, a program is then developed to determine a cure, usually with a diagnostic test (DT).

2.0 TOOLS AND EQUIPMENT

The following equipment will vary, depending on the test selected:

- Multimeter capable of measuring 1 mV_{DC} to 40 V_{DC}, complete with leads with insulated probes
- Copper-copper sulfate reference electrode
- Isolation checker
- Soil resistivity meter, complete with wires and four pins
- Multimeter, including alternating current (AC)/direct current (DC) volts and an ohmmeter
- Current interrupter
- DC ammeter sized for test current
- Battery and control resistor or portable controlled DC power supply
- Pipe locator transmitter and receiver
- Test wires as necessary for the applicable test
- Small hand tools

3.0 SAFETY EQUIPMENT

- Standard safety equipment and clothing, as required by the company's safety manual and regulations
- Electrically insulated clips and probe handles for meter leads
- Only personnel who have received training and are qualified in accordance with local codes and regulations are to work on DC power sources or their supply.

4.0 PRECAUTIONS

The following precautions are in addition to those that must also be followed when working on a particular facility.

4.1 Measure the voltage between the rectifier case and ground before touching the case.
4.2 Open the case expecting to find biting insects, rodents, or snakes inside and take the appropriate precautions.
4.3 Inspect the rectifier for abnormal sounds, temperature, or odors and, if noted, turn it off.
4.4 Switch off the AC voltage supply before installing a current interrupter or each time the taps are adjusted.
4.5 Secure any exposed electrical terminals in a locked container when the rectifier is not attended.
4.6 Measure a structure AC voltage to ground on the structure before taking CP measurements. If the AC voltage to ground is equal to or exceeds 15 V_{AC}, practice the safety measures detailed in NACE SP0177[7.1] and advise other personnel working on the structure of the hazard.
4.7 When working near high-voltage AC (HVAC) power lines, take AC structure-to-ground voltage readings at frequent intervals because these voltages can change with the power line load and geometry.
4.8 Do not work on the structure when lightning is in the area.
4.9 When working near a fence, confirm that it is not an electric fence for livestock (look for insulators) and that an AC voltage is not being induced on it by a parallel HVAC power line.

5.0 PROCEDURE

5.1 Information Required Prior to an Adjustive Survey

5.1.1 Previous annual survey data
5.1.2 Rectifier monitoring and annual data
5.1.3 Drawings
  5.1.3.1 Structure details
  5.1.3.2 CP installation details and location
  5.1.3.3 Test station types and locations
  5.1.3.4 Bond details and locations
5.1.4 Hazardous AC voltage tests
5.1.5 DC interference test results
5.1.6 Critical bond data
5.1.7 Isolation information
5.1.8 Road and railroad casing data (if applicable)
5.1.9 Close interval potential survey data (if applicable)
5.1.10 Pipeline and coating information
5.1.11 In-line inspection or other inspection results (if applicable)
5.1.12 Monthly monitoring data

5.2 Direct Current Power Source

5.2.1 Measure the DC voltage and current output with a portable meter and compare to the panel meter readings.

5.2.2 Turn the power source off and measure the DC voltage between the structure and the anode(s). This will determine the effect of any back electromotive force (EMF) from a galvanic potential difference between the structure metal and anode or carbon in the coke breeze, if applicable, that opposes the rectifier DC voltage. The equivalent circuit in Figure 5.1 demonstrates why this potential opposes the driving potential from the DC power source and must be considered when calculating the external circuit resistance.

5.2.3 Measure the AC voltage between transformer secondary taps (see Figure 5.2). The secondary AC voltage between the coarse and fine tap bars in a rectifier should be slightly more than the DC voltage output.

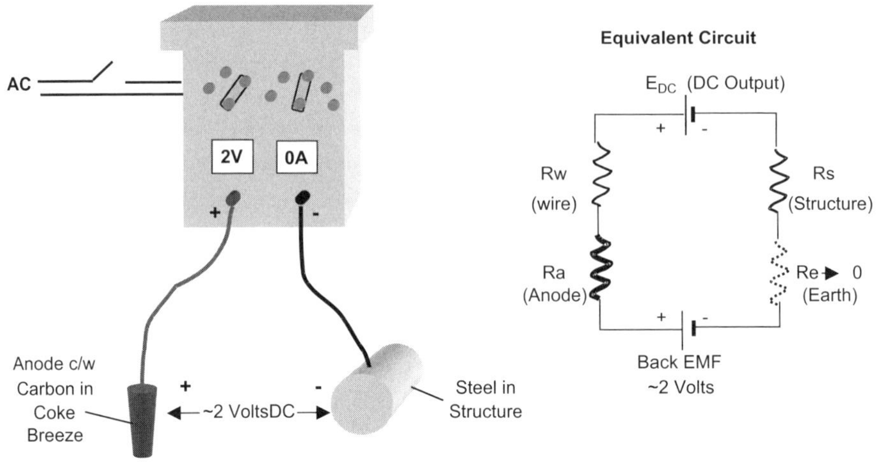

Figure 5.1 Back EMF between structure metal and carbon in coke breeze (AC power source off at left) is shown.

5.2.4 Record the nameplate data, tap setting, voltage, and current output of existing impressed current DC power sources and circuits in both the as-found and test conditions.

5.2.5 If a kilowatt-hour (KWH) meter is available, record the KWH value, the Kh factor, and the revolutions per minute.

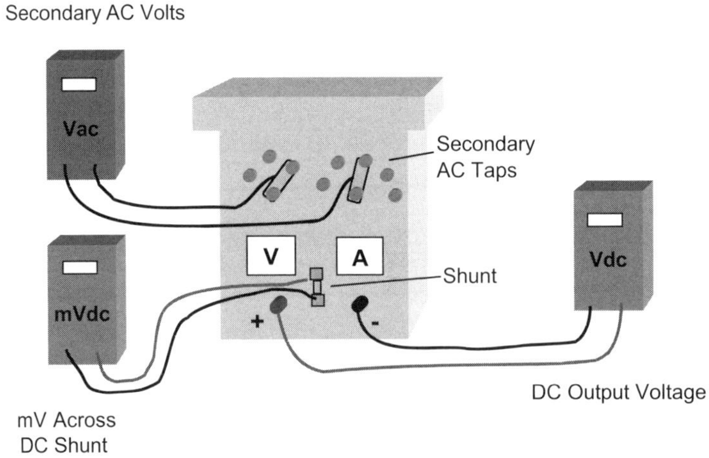

Figure 5.2 Rectifier tests are shown.

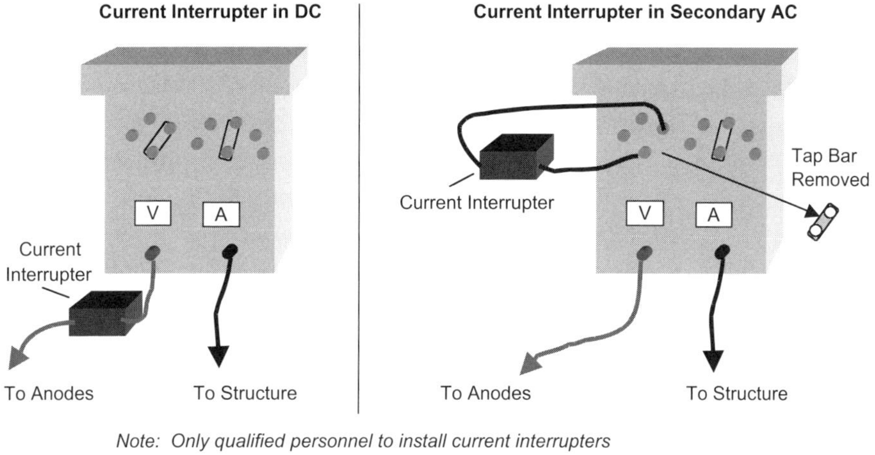

Figure 5.3 Typical current interrupter installations are shown.

5.2.6 For sacrificial anode systems, record the size of the shunt and the millivolt reading across it, and calculate the current output.(1)

5.2.7 Whenever practical, install a current interrupter in either the AC supply, the AC secondary taps, or the DC output of a rectifier, as shown in Figure 5.3, or in the DC output of other impressed current DC power sources. Install the interrupter in series with the sacrificial anodes and any bonds.

5.2.8 For more than one current source, use synchronized interrupters, preferably Global Positioning System time-synchronized interrupters.

5.2.9 Select a long ON and a short OFF cycle to minimize the loss of polarization during the period of interruption, and record the timing of the cycles.

5.2.10 If the feature exists, program the interrupters to turn off after the survey day ends and start again just before it begins to further maintain polarization.

5.3 Structure-to-Electrolyte Potentials

5.3.1 Where practical, measure ON/OFF structure-to-electrolyte potentials, with all influencing DC power sources being

[1] *See* Cathodic Protection Procedure No. 3: Direct Current Measurements, Section 6.3.

interrupted, at available contact points to the structure to determine if the criterion for CP is being met.[7.1(2)]

5.3.2 Use a high-input impedance voltmeter (10 MΩ minimum) in conjunction with a copper-copper sulfate ($Cu/CuSO_4$) reference electrode (CSE) for soil or freshwater environments or a silver-silver chloride reference electrode (SCC) for high-brine conditions.

5.3.3 Calibrate the field CSE by measuring a potential to a new, clean standard CSE that was recently charged with distilled water and copper sulfate crystals. Similarly, measure a saturated SCC to a new, clean SCC. Calibrate an open or seawater SCC in seawater.[(3)] Replace the reference electrode if the potential difference is still greater than 5 mV.

5.3.4 Connecting the voltmeter positive to the structure and the negative to the reference electrode is now the preferred method. (*Do not* apply this polarity when connecting a DC power source to a structure.) When connected in this manner, structure-to-electrolyte potential readings should then be negative and recorded as such. Connecting the voltmeter with the voltmeter negative to the structure is still permissible; however, the tester must realize that when a positive structure-to-electrolyte potential value is obtained, the reading is negative and must be recorded as such.

5.3.5 In rocky, sandy, very dry soils or frozen ground, add water to the ground surface or a damp sponge attached to the reference electrode. In extreme conditions, a multiple-input impedance interface or multi-input impedance meters may be used. The potential measurements at a minimum of two input impedances must be the same; otherwise, the reference cell circuit resistance must be further reduced.

5.3.6 Document all techniques and raw data used to improve the quality of the data.

5.3.7 Determine the frequency of structure-to-electrolyte potential measurements.

---

[2] *See* Cathodic Protection Procedure No. 2: Structure-to-Electrolyte Potential Measurement.

[3] *See* Cathodic Protection Procedure No. 2: Structure-to-Electrolyte Potential Measurement, Appendix A.

5.3.8 Interrupt all influencing DC current sources on a recorded timed ON and OFF cycle.

5.3.9 When possible, install a stationary data logger to witness the interruption cycles and to confirm that all interrupters continue to operate and remain in synchronization. The stationary recorder will also serve to verify synchronized interruption and if depolarization occurred during the survey interruption period.

5.3.10 Take instant OFF structure-to-electrolyte potentials with all influencing DC power sources interrupted wherever practical.

5.3.11 Record the instant OFF potential reading between 0.6 and 1.0 s after interruption. If using a digital voltmeter, record the second reading displayed after interruption as the first display may be an average of the dropping values from the ON potential.

5.3.12 For pipelines, the spacing of the test station structure-to-electrolyte (pipe-to-soil) potential measurements should be approximately 3 km (2 mi); however, closer readings will allow a more accurate analysis of the data.

5.3.13 Obtain a complete set of data, which may include the following:

5.3.13.1 ON/OFF structure-to-electrolyte potential tests with all influencing current sources being interrupted to relate to the polarized potential criterion. All current sources include DC power sources (rectifiers; thermoelectric generators; and solar, wind, and engine generators), sacrificial anodes, and bonds.

5.3.13.2 ON structure-to-electrolyte potential tests where current sources cannot be interrupted. In this case, additional testing must be completed to predict the *IR* drop error.[(4)]

5.3.13.3 Depolarization potential tests (preferably a CIS), in which the current is left off after an ON/OFF potential test is advisable

5.3.13.4 DC current source outputs and bond current data

5.3.13.5 Measurement of structure-to-electrolyte potentials on each side of isolating features,[(5)] on foreign structures, and on road or railroad casings[(6)]

---

[4] *See* Cathodic Protection Procedure No. 2: Structure-to-Electrolyte Potential Measurement.

[5] *See* Cathodic Protection Procedure No. 9: Electrical Isolation.

[6] *See* Cathodic Protection Procedure No. 10: Road and Railroad Cased Crossings (Basic).

5.3.13.6 AC structure-to-ground voltages, to be measured prior to the DC structure-to-electrolyte potentials

5.3.13.7 Comparison of the survey data to the last survey results to confirm that the location being tested is the intended area and that the operation of the CP system is similar

5.3.14 ON Structure-to-Electrolyte Potentials

5.3.14.1 Determine the frequency of structure-to-electrolyte potential measurements.

5.3.14.2 Measure ON structure-to-electrolyte potentials and identify each location with a stake or paint so that the reference electrode can be placed in the exact position for subsequent tests.

5.3.14.3 Determine the *IR* drop component that is included in each ON structure-to-electrolyte potential.[7]

5.3.14.4 Calculate the true polarized potential by removing the *IR* drop error from the ON potential that was measured.

5.3.15 ON/OFF Structure-to-Electrolyte Potentials

5.3.15.1 Ensure that all influencing DC power sources are being interrupted, including all sources of CP and bonds to the structure. To determine the influence of a DC power source or bond, interrupt it alone and note the change in potential through the area under testing. Alternately, interrupt a series of DC power sources on different identifiable cycles and record the effect of each through the area under testing.

5.3.15.2 Record the ON potential just before interruption and the OFF potential just after interruption. If using a digital meter, record the second clear detectable reading after interruption.[8]

5.3.15.3 If using a data logger, take the reading that is more than 0.6 s after interruption to avoid capturing a spike during the interruption.

---

[7] *See* Cathodic Protection Procedure No. 2: Structure-to-Electrolyte Potential Measurement.

[8] *See* Cathodic Protection Procedure No. 2: Structure-to-Electrolyte Potential Measurement.

5.3.16 Depolarization Potentials

5.3.16.1 Determine the frequency of structure-to-electrolyte potential measurements.

5.3.16.2 Interrupt all influencing DC current sources on an ON and OFF cycle and record the time for each cycle.

5.3.16.3 Measure ON/OFF structure-to-electrolyte potentials and identify each location with a stake or paint so that the reference electrode can be placed in the exact position for subsequent depolarized potential tests.

5.3.16.4 Turn off all current sources and record spot structure-to-electrolyte potentials over time, until the potentials have become relatively stable. In some cases, this may take up to several days. A stationary data logger will facilitate this test.

5.3.16.5 When the potentials have stabilized, measure the depolarized structure-to-electrolyte potential with the reference electrode placed in the same exact locations as during the ON/OFF structure-to-electrolyte potential survey.

5.3.16.6 Calculate the depolarization (polarization) at each measurement location using Equation (5.1):

$$\Delta V_p = E_{\text{off}} - E_{\text{depol}}, \tag{5.1}$$

where

ΔV_p depolarization (polarization) (volts)

E_{off} instant OFF structure-to-electrolyte potential (volts)

E_{depol} depolarized structure-to-electrolyte potential (volts)

5.4 Direct Current Stray Current Tests (Interference)

5.4.1 Determine if the structure under testing is being affected by a dynamic DC stray current.(9)

9 *See* Cathodic Protection Procedure No. 8: Direct Current Stray Current Interference.

5.4.2 Measure foreign structure-to-electrolyte potentials where they may be affected by the CP installations. Where the foreign facilities are affected, make a sketch of the configuration of the foreign facilities, in addition to the data.

5.4.3 Should no test station exist for the foreign structure, with the owner's permission, test other locations to note the effect of the proposed CP system. Make sketches of the foreign structure and reading locations.

5.4.4 Existence of bonds with foreign structures should be recorded, including current and its direction.

5.4.5 Verify the accuracy of adjacent potentials that differ in excess of 20%. Retest the unusual potential changes immediately to confirm proper reference electrode contact with the ground by the addition of water or by exposing moist soil. This change may be due to a slow telluric current variation.

5.4.6 Telluric or other dynamic stray current activity requiring calibration can be defined as an OFF potential fluctuation exceeding 20 mV peak to peak over the duration of the testing.

5.4.7 If a telluric or another dynamic stray current is detected, install data loggers at the one-fourth and three-fourths points of the section to be tested to record structure-to-electrolyte (pipe-to-soil) potentials versus time and leave them recording for a period of approximately 22–24 hours, where practical.

5.4.8 Alternately, measure structure-to-electrolyte potentials manually and record the values and the time of each reading. Plot the results to see any trends.

5.4.9 If the test section is less than 1.6 km (1.0 mi), a single data logger may be installed at the test site.

5.4.10 Record each test station structure-to-electrolyte potential with another data logger for a period of 5 min.

5.5 Investigate the Cause of Poor Cathodic Protection

5.5.1 When practical, identify the source of any problems encountered during the adjustive survey.[(10)]

---

[10] *See* Cathodic Protection Procedure No. 4: Diagnostic Testing (Current Requirement).

5.5.2 Inspect the DC power sources and compare with the target DC volts and DC amperes. If the target outputs are significantly different from the target, complete the following inspections:

5.5.2.1 If there is 0 to ~ 2 V_{DC}[11] and 0 A_{DC} output, look for trouble in the rectifier or the AC supply to the rectifier. Confirm that the ~2 V_{DC} reading is from the DC power source by turning it off and disconnecting one DC cable. If the reading stays at ~2 V_{DC}, then it is from the DC power source. If it drops to 0 V, then it is the galvanic difference between the structure material and anode or the carbon in the coke breeze.

5.5.2.2 If there is a normal voltage output but 0 A_{DC} output, look for the trouble in the cables, anodes, or connections external to the rectifier.

5.5.2.3 If there is approximately one-half normal voltage and approximately one-half normal current, then investigate the possibility of a failed diode, causing the rectifier to half-wave.

5.5.2.4 If a problem is suspected in the rectifier or the external DC circuit, refer to "Cathodic Protection Procedure No. 1: Rectifier Adjustment, Inspection, and Basic Troubleshooting." Complete repairs before proceeding with the adjustive survey.

5.5.3 Inspect the DC bonds and repair any broken bonds found.

5.5.4 Test all isolating features.[12]

5.5.5 Test all road or railroad casings, if applicable, to confirm that they are isolated.[13]

5.5.6 If the problems causing the loss of CP when corrected restore protection, then the adjustive survey is completed.

5.5.7 When the preceding faults are corrected but a CP criterion has still not been met, proceed with a DT[14] if CP does not return.

---

11 The 2 V_{DC} may be the galvanic difference between the steel in the structure and the carbon in the coke breeze.

12 *See* Cathodic Protection Procedure No. 9. Electrical Isolation.

13 *See* Cathodic Protection Procedure No. 10: Road and Railroad Cased Crossings (Basic).

14 *See* Cathodic Protection Procedure No. 4: Diagnostic Testing (Current Requirement).

5.6 Pipeline Current Measurements

Where practical, measure the current in a pipeline at intervals to determine the distribution of current with the test current both OFF and ON.[(15)]

5.7 Alternating Current Voltage on Structures

5.7.1 Measure the AC structure-to-ground voltage before taking a DC structure-to-electrolyte potential to confirm that a hazardous voltage does not exist.

5.7.2 If an AC structure-to-ground voltage of 15 V_{AC} or greater exists, then follow the guidelines in NACE SP0177[7.1] and advise other personnel that the hazardous conditions exist.

6.0 ANALYSIS

6.1 Criteria to Be Met

6.1.1 CP criteria are detailed in Section 6.0 of NACE SP0169.[7.2] Similar criteria are given in CGA OCC-1[7.3] and ISO 15589-1.[7.4] [(16)] Test procedures to determine these criteria are given in NACE TM0497.[7.5]

6.1.2 There are three structure-to-electrolyte potential criteria for submerged or buried steel structures in the absence of specific data that demonstrate that adequate CP has been applied, including the following:

6.1.2.1 A negative (cathodic) potential of at least 850 mV (with respect to a CSE contacting the electrolyte) with the cathodic protection applied but with voltage drops other than across the structure-to-electrolyte interface (*IR* drop) removed. *IR* drops between the reference electrode and the structure-to-electrolyte boundary are an error in this reading and must be removed from the ON potential before applying to this criterion, as illustrated by Equation (6.1). NACE SP0169[7.1] discusses methods to evaluate the

---

15 *See* Cathodic Protection Procedure No. 3: Direct Current Measurements.

16 Use latest version. Note ISO 15589-1 includes only polarized potential and polarization criteria.

IR drop, which is essentially the difference between the ON potential and the instant OFF potential:

$$E_c = (E_{\text{on}} - IR), \tag{6.1}$$

where

- E_c potential for criterion (-850 mV_{CSE} or more negative)
- E_{on} potential with current applied (millivolts)
- IR voltage drop between the reference electrode and the structure-to-electrolyte boundary (millivolts)

6.1.2.2 A negative polarized potential equal to or more electronegative than -850 mV relative to a saturated copper-copper sulfate reference electrode contacting the electrolyte. This can be obtained by interrupting all current sources influencing the structure and taking an instant OFF potential.

6.1.2.3 A minimum of 100 mV of cathodic polarization between the structure surface and a stable reference electrode contacting the electrolyte. Polarization is the change in potential from the native or free corroding potential and the instant OFF potential, as illustrated by Equation (6.2). It can be measured during formation or decay after all influencing current sources have been turned off:

- Polarization:

$$\Delta E_p = (E_{\text{off}} - E_{\text{native}}) \tag{6.2}$$

- Depolarization

$$\Delta E_{\text{depol}} = (E_{\text{off}} - E_{\text{depol}}), \tag{6.3}$$

where

- ΔE_p polarization for criterion (100-mV minimum)

E_{off}	potential with all current momentarily interrupted (millivolts)
E_{native}	native (free corroding) potential before CP current is applied (millivolts)
ΔE_{depol}	depolarization for criterion (100-mV minimum)
E_{depol}	depolarized potential with current left off (millivolts)

6.1.3 Only one of these criteria needs to be met. For example, if a polarized potential is less negative than -850 mV_{CSE}, a depolarization survey may prove that the 100-mV criterion is being achieved.

6.2 Cause of Subcriterion Potentials

6.2.1 Some common causes of a system's inability to meet criteria, tests, and cures are listed in Table 6.1.

6.2.2 Compare the DC power output to the historical information. If they are similar, review additional data.

6.3 Structure-to-Electrolyte Potentials

6.3.1 Confirm that all structure-to-electrolyte potentials meet one of the criteria for CP. Where readings do not meet this criterion, complete the subsequent analysis.

6.3.2 Compare the structure-to-electrolyte potentials with the historical data.

6.3.3 Readings that are more electronegative than before at the same current output may suggest that the structure has been reduced in size, either intentionally or unintentionally, such as by a faulty bond isolating part of the structure, or that a CP system that is at a more electronegative potential is now shorted to the structure.

6.3.4 Readings that are more electronegative than before at the same current output may also suggest anodic interference; that is, a ground or anode bed of a foreign DC power source is in close proximity to the structure.

Table 6.1 Summary of Identification of CP Trouble, Tests, and Causes

System	DC Power Source		Secondary	Structure-to-Electrolyte	Trouble		
Component	Volts	Amperes	AC Voltage	Potentials	Suspected	Tests	Remedy
Sacrificial Anode							
		A–		P–	May not be trouble. Anode current will decrease when potential becomes more negative.	No action	
		A+		P+	Opposite to above	Test for shorts or faults in the structure system.	Repair as required.
		A–		P+	Anodes failing	Measure anode-to-electrolyte potential. Perform anode voltage gradient test.	Replace anodes as necessary.
DC Power Source							
	0 to ~2*	0	0	P+	No AC power or failed DC power components. Trouble is before or in DC power source.	Confirm AC supply, check circuit breaker, test fuses, poor connections, or broken wires in DC power source. Check for signs of heat. If a battery supplemented source, test batteries for charge.	Confirm cause and correct before re-energizing
						If circuit breaker trips, look for short.	If no short, reduce DC output voltage

(continued)

Table 6.1 *(continued)*

System Component	DC Power Source		Secondary AC Voltage	Structure-to-Electrolyte Potentials	Trouble Suspected	Tests	Remedy
	Volts	Amperes					
	0 to ~2*	0	0	P+	No secondary AC voltage at taps	Test AC supply and circuit breaker.	If AC supply and circuit breaker are OK, then test transformer.
	0 to ~2*	0	V	P+	Fuse(s) in DC power source, failed rectifying element, poor connections or faulty wire	Test fuse(s), rectifying element, connections, and wires.	If fuse, test for short. If none found, lower voltage and re-energize. Otherwise, replace or repair as required.
	~ 1/2 V	~ 1/2 A	V	P+	One-half wave DC output. One part of rectifier bridge circuit is open.	Turn off, remove rectifying element connections, and test each diode or element.	Replace rectifying element(s).
	?V	?A	V	P	Faulty meters	Calibrate meters.	Replace as necessary.
DC Cables/Anode Bed							
	V	0	V	P+	Faulty cable, connections, or anodes	Trace cable to structure and anodes. Perform anode voltage gradient test.	Repair or replace as required.

	V	Dropping over time	V	P+	Failing or dry anodes	Potential profile over anodes to confirm status. Turn off to note recovery.	Temporary cure may be to water anodes. Replace anodes as required.
Structure							
	V	A	V	P+	Shorted isolation, accidental contact to foreign structure, faulty bonds, deteriorating coating	Test isolation and bonds. Trace for contacts. Complete DT test.	Repair isolation or bonds. Separate any contacts to foreign structures. Add CP capacity for poor coating or recoat.

Legend:
V Normal voltage
A Normal current
P Normal structure-to-electrolyte potential
V+ Greater than normal voltage
V− Lower than normal voltage
A+ Greater than normal current
A− Lower than normal current
?V, ?A Abnormal readings or varying
P+ Structure-to-electrolyte potential more electropositive
P− Structure-to-electrolyte potential more electronegative
* ~2 V may be due to galvanic difference between steel and anode or carbon in the coke breeze and not an indication of power

6.3.5 Readings that are more electropositive than before may indicate one or more of the following problems:

6.3.5.1 Low DC output from the DC power source

6.3.5.1.1 Failing anodes

6.3.5.1.2 No AC power or energy source to the DC power supply

6.3.5.1.3 Faulty components in the DC power source

6.3.5.1.4 Faulty cables or connections in the external DC circuit

6.3.5.1.5 A short in the AC or DC circuitry

6.3.5.2 Shorted isolation, adding a foreign structure to the CP system and increasing the current requirements if the foreign system is at more electropositive potentials

6.3.5.3 Shorted casings, adding more bare metal to the structure and, therefore, increasing the current required for protection

6.3.5.4 A contact to a foreign structure at a less electronegative potential, causing an electropositive shift in the structure potential

6.3.5.5 Deteriorating coating, resulting in an increase in the current necessary to achieve criteria

6.3.5.6 A faulty interference bond or new interference from a foreign DC power source

6.4 Dynamic Stray Current

6.4.1 If telluric or other dynamic stray currents are detected, one method of correction is described subsequently. Data loggers are to be installed near each end of the section to be tested to record pipe-to-electrolyte (pipe-to-soil) potentials versus time and are to be left recording for a period of approximately 22 hours, where practical.

6.4.2 If the test section is less than 1.6 km (1.0 mi), a single data logger may be installed at the test site.

6.4.3 Each test station pipe-to-electrolyte potential is to be recorded with another data logger for a period of 5 min; otherwise, CIS potentials are taken as normal.

6.4.4 Telluric or other dynamic stray current activity requiring calibration is defined as an OFF potential fluctuation exceeding 20 mV peak to peak over the duration of the testing.

6.4.5 The true potential at the stationary data loggers shall first be determined either by a quiet period or by an average over the test period. For each potential measurement along the line, the difference in potential between that and the stationary data logger at the same moment in time is to be determined. This difference added to the difference to the true potential at the stationary data logger is the correction factor for the portable data logger reading. Equations (6.4)–(6.6) are presented for the following situations:

- If two stationary data loggers are used, the method given in Equations (6.4) and (6.5) applies:

$$\acute{\varepsilon}_b = [\acute{\varepsilon}_a(c - b)/c] + [\acute{\varepsilon}_c(b - a)/c], \tag{6.4}$$

where

a	first stationary potential location (volts)
b	portable potential location (volts)
c	second stationary potential location (volts)
$\acute{\varepsilon}_a$	error in potential at a at time x (volts)
$\acute{\varepsilon}_b$	error in potential at b at time x (volts)
$\acute{\varepsilon}_c$	error in potential at c at time x (volts)

and

$$E_p = E_{p\,\text{measured}} - \acute{\varepsilon}_b, \tag{6.5}$$

where

E_p	true potential at the portable data logger location (volts)
$E_{p\,\text{measured}}$	potential at the portable data logger location (volts)

- If only one stationary data logger is used, then Equation (6.6) applies:

$$E_p = E_s - (E_{sa} - E_{pa}), \tag{6.6}$$

where

E_p true potential at the portable data logger location
E_s true potential at the stationary location
E_{sa} stationary potential at time *a* during the data logging
E_{pa} portable potential at time *a* during the data logging

6.4.6 Other methods can also be used to correct for dynamic stray currents.

6.4.7 The stationary recorder will also serve to verify synchronized interruption and if depolarization occurred during the survey interruption period.

6.5 Direct Current Power Source Interruption

6.5.1 Review stationary data log profiles to confirm that the interrupters at all influencing rectifiers continued to function during the test.

6.5.2 Identify any structure-to-electrolyte potential readings that may have been affected by an interrupter malfunction.

6.5.3 Note the amount of depolarization that took place during the interruption period.

6.6 Pipeline Current Measurement

6.6.1 Current Span Method

6.6.1.1 Calculate the pipeline current, where possible.

6.6.1.2 Calculate the current pickup and discharge for each pipeline section.

6.6.2 Clamp-On Ammeter

Calculate the current pickup and discharge for each pipeline section.

6.7 Rectifier Efficiency

Calculate the rectifier efficiency(17) and compare it with historical data.

---

17 *See* Cathodic Protection Procedure No. 1: Rectifier Measurement, Inspection, and Basic Troubleshooting.

7.0 REFERENCES

7.1 NACE Standard SP0177-2007, "Mitigation of Alternating Current and Lightning Effects on Metallic Structures and Corrosion Control Systems" (Houston, TX: NACE International, 2007).

7.2 NACE Standard SP0169-2007, "Control of External Corrosion on Underground or Submerged Metallic Piping Systems" (Houston, TX: NACE International, 2007).

7.3 CGA Recommended Practice OCC-1-2005, "Control of Corrosion on Buried or Submerged Metallic Piping Systems" (Canadian Gas Association, Ottawa, Ontario, 2005).

7.4 ISO 15589-1, "Petroleum and Natural Gas Industries: Cathodic Protection of Pipeline Transportation Systems, Part 1—On-Land Pipelines" (Geneva, Switzerland: ISO, 2003).

7.5 NACE Standard TM0497-2002, "Measurement Techniques Related to Criteria for Cathodic Protection on Underground or Submerged Metallic Piping Systems" (Houston, TX: NACE International, 2002).

CATHODIC PROTECTION PROCEDURES NO. 6

Commissioning Survey 6

1.0 INTRODUCTION

The purpose of the commissioning survey is to confirm that the new cathodic protection (CP) equipment meets the design specifications to energize the CP system and to adjust it to achieve a CP criterion. If this cannot be met within the design capacity of the CP system, a program is then developed to determine a cure.

Although not called "commissioning surveys," some of the surveys are described in the literature.[7.1–7.6]

2.0 TOOLS AND EQUIPMENT

The following equipment will vary, depending on the test selected:

- Multimeter capable of measuring 1 mV_{DC} to 40 V_{DC}, complete with leads with insulated probes
- Copper-copper sulfate reference electrode
- Isolation checker
- Soil resistivity meter, complete with wires and four pins
- Multimeter including alternating current (AC)/direct current (DC) volts and an ohmmeter
- Current interrupter
- DC ammeter sized for test current
- Battery and control resistor or portable controlled DC power supply

- Pipe locator transmitter and receiver
- Test wires as necessary for the applicable test
- Small hand tools

3.0 SAFETY EQUIPMENT

- Standard safety equipment and clothing, as required by the company's safety manual and regulations
- Electrically insulated clips and probe handles for meter leads
- Only personnel who have received training and are qualified in accordance with local codes and regulations are to work on DC power sources or their supply

4.0 PRECAUTIONS

The following precautions are in addition to those that must also be followed when working on a particular facility.

4.1 Determine the location of all AC power supply disconnects.

4.2 Measure the voltage between the rectifier case and ground before touching the case and immediately after energizing.

4.3 Open the case expecting to find biting insects, rodents, or snakes inside and take the appropriate precautions.

4.4 Inspect the rectifier for abnormal sounds, temperature, or odors and, if noted, turn it off.

4.5 Switch off the AC voltage supply before installing a current interrupter or each time the taps are adjusted.

4.6 Secure any exposed electrical terminals in a locked container when the rectifier is not attended.

4.7 Measure a structure AC voltage to ground on the structure before taking CP measurements. If the AC voltage to ground is equal to or exceeds 15 V_{AC}, practice the safety measures detailed in NACE SP0177[7.1] and advise other personnel working on the structure of the hazard.

4.8 When working near high-voltage AC (HVAC) power lines, take AC structure-to-ground voltage readings at frequent intervals as these voltages can change with the power line load and geometry.

4.9 Do not work on the structure when lightning is in the area.

4.10 When working near a fence, confirm that it is not an electric fence for livestock (look for insulators) and that an AC voltage is not being induced on it by a parallel HVAC power line.

5.0 PROCEDURE

5.1 Information Required Prior to a Commissioning Survey

5.1.1 Design information

5.1.2 Drawings

5.1.2.1 Structure details

5.1.2.2 CP installation details and location

5.1.2.3 Test station types and locations

5.1.2.4 Bond details and locations

5.1.3 Hazardous AC voltage expectations

5.1.4 DC interference expectations in the design

5.1.5 Isolation information

5.1.6 Road and railroad casing data (if applicable)

5.1.7 Pipeline and coating information

5.2 Direct Current Power Source

5.2.1 Only those qualified are to complete tests on the DC power sources.

5.2.2 Record the nameplate data of the DC power supply.

5.2.3 Transformer Secondary Tap Rectifier

5.2.3.1 Measure the AC supply and confirm that it agrees with the AC voltage rating of the rectifier. If the rectifier is a dual–AC voltage input, confirm that the AC input taps are set to those of the AC voltage supply. Do not energize the rectifier until the AC supply matches the rectifier rating.

5.2.3.2 Measure the tap-to-tap AC voltage between transformer secondary taps (see Figure 5.1). The fine tap-to-tap AC voltages should be about equal and sum to that of one coarse tap. The coarse taps should also be about the same AC voltage.

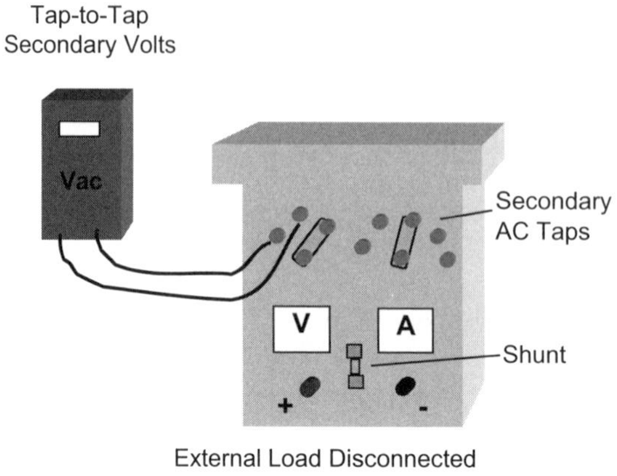

Figure 5.1 Measuring secondary tap-to-tap AC voltage is shown.

5.2.3.3 Measure the DC voltage between the structure and the anode(s). This will determine the effect of any back electromotive force (EMF) from a galvanic potential difference between the structure metal and anode or carbon in the coke breeze, if applicable, that opposes the rectifier DC voltage (Figure 5.2).

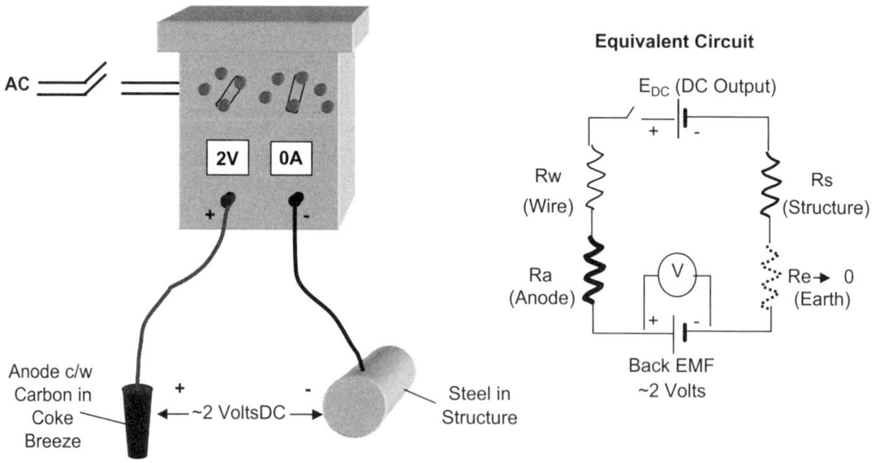

Figure 5.2 Back EMF between structure metal and carbon in coke breeze is shown. AC power source is off.

5.2.3.4 To test for the external circuit resistance, adjust the taps such that the secondary AC voltage exceeds this value by 2–4 V_{AC}. Energize and measure the DC volts and DC amperes. Calculate the external circuit resistance using Equation (5.1):

$$R_a = \frac{(E_{DC1} - \text{Back EMF})}{I_{DC1}}, \tag{5.1}$$

where

R_a	resistance of anode circuit (ohms)
E_{DC1}	test DC voltage output (volts)
Back EMF	DC volts between structure and anode before energizing (volts)
I_{DC1}	test DC current output (amperes)

5.2.3.5 Calculate the DC voltage required for the design current using Equation (5.2):

$$E_{DC} = I_{DC} R_a + \text{Back EMF}, \tag{5.2}$$

where

E_{DC}	DC voltage output required (volts)
I_{DC}	design DC current output (amperes)
R_a	resistance of anode circuit (ohms)
Back EMF	DC volts between structure and anode before energizing (volts)

5.2.3.6 Set the taps to approximately 10% to 15% more than the preceding E_{DC} value. The secondary AC voltage between the coarse and fine tap bars in a rectifier will normally be approximately 15% greater than the DC voltage output; however, this will vary between rectifiers and the percentages of the total rating at which the rectifier is operating. Either refer to the preceding measurements or measure AC volts between the coarse and fine taps to determine the actual setting.

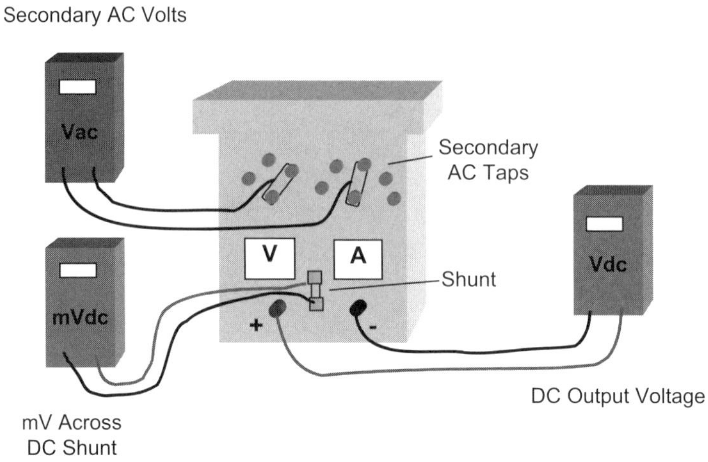

Figure 5.3 Rectifier tests are shown.

5.2.3.7 Energize the rectifier and measure the DC voltage and current outputs. Adjust the taps to meet the required current output.

5.2.3.8 Measure the DC voltage and current output with a portable meter and compare to the panel meter readings, as shown in Figure 5.3.

5.2.4 Record the nameplate data, tap setting, voltage, and current output of existing impressed current DC power sources and circuits in both the as-found and test conditions.

5.2.5 For sacrificial anode systems, record the size of the shunt, the millivolt reading across it, and the current output.

5.2.6 Whenever practical, install a current interrupter in the AC supply, the AC secondary taps, or the DC output of a rectifier, as shown in Figure 5.4, or in the DC output of other impressed current DC power sources. Install the interrupter in series with the sacrificial anodes and any bonds.

5.2.7 For more than one current source, use synchronized interrupters, preferably Global Positioning System time-synchronized interrupters.

5.2.8 Select a long ON and a short OFF cycle to minimize the loss of polarization during the period of interruption and record the timing of the cycles.

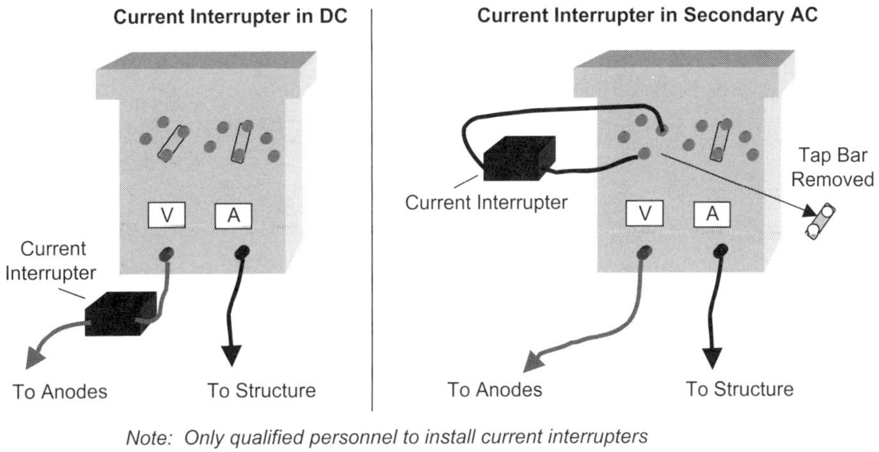

Figure 5.4 Typical current interrupter installations are shown.

5.2.9 If the feature exists, program the interrupters to turn off after the survey day ends and start again just before it begins to further maintain polarization.

5.3 Structure-to-Electrolyte Potentials

5.3.1 Where practical, measure ON/OFF structure-to-electrolyte potentials, with all influencing DC power sources being interrupted, at available contact points to the structure to determine if the criterion for CP is being met.[7.1 (1)]

5.3.2 Use a high-input impedance voltmeter (10 MΩ minimum) in conjunction with a copper-copper sulfate ($Cu/CuSO_4$) reference electrode (CSE) for soil or freshwater environments or a silver-silver chloride reference electrode (SCE) for high-brine conditions.

5.3.3 Calibrate the field CSE by measuring a potential to a new, clean standard CSE that was recently charged with distilled water and copper sulfate crystals.[(2)] Similarly, measure a saturated SCE to a

---

[1] *See* Cathodic Protection Procedure No. 2: Structure-to-Electrolyte Potential Measurement.

[2] *See* Cathodic Protection Procedure No. 2: Structure-to-Electrolyte Potential Measurement, Appendix A.

new, clean SCE. Calibrate an open or seawater SCE in seawater. Replace the reference electrode if the potential difference is still greater than 5 mV.

5.3.4 Connecting the voltmeter positive to the structure and the negative to the reference electrode is now the preferred method. (*Do not* apply this polarity when connecting a DC power supply.) When connected in this manner, structure-to-electrolyte potential readings should then be negative and recorded as such. Connecting the voltmeter with the voltmeter negative to the structure is still permissible; however, the tester must realize that the leads are reversed, and, thus, when a positive structure-to-electrolyte potential value is obtained, the reading is negative and must be recorded as such.

5.3.5 In rocky, sandy, very dry soils or frozen ground, add water to the ground surface or a damp sponge attached to the reference electrode. In extreme conditions, a multiple-input impedance interface or multi-input impedance meters may be used. The potential measurements at a minimum of two input impedances must be the same; otherwise, the reference cell circuit resistance must be further reduced.

5.3.6 Document all techniques and raw data used to improve the quality of the data.

5.3.7 Determine the frequency of structure-to-electrolyte potential measurements.

5.3.7 For pipelines, the spacing of the test station structure-to-electrolyte (pipe-to-soil) potential measurements should be approximately 3 km (2 mi); however, closer readings will allow a more accurate analysis of the data.

5.3.8 Obtain a complete set of data, which may include the following:

5.3.8.1 ON/OFF structure-to-electrolyte potential tests with all influencing current sources being interrupted to relate to the polarized potential criterion. All current sources include DC power sources (rectifiers; thermoelectric generators; and solar, wind, and engine generators), sacrificial anodes, and bonds.

5.3.8.2 ON structure-to-electrolyte potential tests where current sources cannot be interrupted. In this case, additional testing must be completed to predict the *IR* drop error.(3)

5.3.8.3 Depolarization potential tests (preferably a CIS[4]), where the current is left off after an ON/OFF potential test is advisable

5.3.8.4 DC current source outputs and bond current data

5.3.8.5 Measurement of structure-to-electrolyte potentials on each side of isolating features,(5) on foreign structures, and on road or railroad casings(6)

5.3.8.6 Measurement of AC structure-to-ground voltages prior to the DC structure-to-electrolyte potentials to determine if hazardous voltages are present

5.3.8.7 Comparison of the survey data to the last survey results to confirm that the location being tested is the intended area and that the operation of the CP system is similar

5.3.9 ON Structure-to-Electrolyte Potentials

5.3.9.1 Determine the frequency of structure-to-electrolyte potential measurements.

5.3.9.2 Measure ON structure-to-electrolyte potentials and identify each location with a stake or paint so that the reference electrode can be placed in the exact position for subsequent tests.

5.3.9.3 Determine the *IR* drop component that is included in each ON structure-to-electrolyte potential.(7)

5.3.9.4 Calculate the true polarized potential by removing the *IR* drop error from the ON potential that was measured.

5.3.10 ON/OFF Potentials

5.3.10.1 Interrupt all influencing DC current sources on a recorded timed ON and OFF cycle.

5.3.10.2 When possible, install a stationary data logger to witness the interruption cycles and to confirm that all interrupters continue to operate and remain in

---

[3] *See* Cathodic Protection Procedure No. 2: Structure-to-Electrolyte Potential Measurement.

[4] *See* Cathodic Protection Procedure No. 7: Close Interval Potential Survey.

[5] *See* Cathodic Protection Procedure No. 9: Electrical Isolation.

[6] *See* Cathodic Protection Procedure No. 10: Road and Railroad Cased Crossings (Basic).

[7] *See* Cathodic Protection Procedure No. 2: Structure-to-Electrolyte Potential Measurement.

synchronization. The stationary recorder will also serve to verify synchronized interruption and if depolarization occurred during the survey interruption period.

5.3.10.3 Measure the instant OFF structure-to-electrolyte potentials with all influencing DC power sources interrupted wherever practical. Identify the OFF potentials by the length of the cycle only.

5.3.10.4 Record the instant OFF potential reading between 0.6 and 1.0 s after interruption when using a rapid data collection instrument. If using a digital voltmeter, record the second reading displayed after interruption as the first display may be an average of the dropping values from the ON potential.

5.3.11 Depolarization Potentials

5.3.11.1 Determine the frequency of structure-to-electrolyte potential measurements.

5.3.11.2 Interrupt all influencing DC current sources on an ON and OFF cycle and record the time for each cycle.

5.3.11.3 Measure ON/OFF structure-to-electrolyte potentials and identify each location with a stake or paint so that the reference electrode can be placed in the exact position for subsequent tests.

5.3.11.4 Turn off all current sources and record spot structure-to-electrolyte potentials over time, until the potentials have become relatively stable. In some cases, this may take up to several days. A stationary data logger will facilitate this test.

5.3.11.5 When the potentials have stabilized, measure the depolarized potential with the reference electrode placed in the same exact locations as during the ON/OFF structure-to-electrolyte potential survey.

5.3.11.6 Calculate the depolarization at each measurement location using Equation (5.3):

$$\Delta V_p = E_{\text{off}} - E_{\text{depol}}, \tag{5.3}$$

where

ΔV_p depolarization (polarization) (volts)

E_{off} instant OFF structure-to-electrolyte potential (volts)

E_{depol} depolarized structure-to-electrolyte potential (volts)

5.4 Direct Current Stray Current Tests (Interference)

5.4.1 Determine if the structure under testing is being affected by a dynamic DC stray current.[8]

5.4.2 Measure foreign structure-to-electrolyte potentials where they may be affected by the CP installations. Where the foreign facilities are affected, make a sketch of the configuration of the foreign facilities, in addition to the data.

5.4.3 With the owner's permission, test other locations on foreign systems to note the effect of the proposed CP system. Make sketches of the foreign structure relative to the protected structure, including reading locations.

5.4.4 Record the existence of bonds with foreign structures, including current and its direction.

5.4.5 Verify the accuracy of adjacent potentials that differ in excess of 20%. Test the potential spikes immediately to confirm proper reference electrode contact with the ground by the addition of water or by exposing moist soil. This change may be due to a slow telluric current variation.

5.4.6 Telluric or other dynamic stray current activity requiring calibration can be defined as an OFF potential fluctuation exceeding 20 mV peak to peak over the duration of the testing.

5.4.7 If a telluric or another dynamic stray current is detected, install data loggers at the one-fourth and three-fourths points of the section to be tested to record structure-to-electrolyte (pipe-to-soil) potentials versus time and leave them recording for a period of approximately 22–24 hours, where practical.

---

[8] *See* Cathodic Protection Procedure No. 8: Direct Current Stray Current Interference.

5.4.8 Alternately, measure structure-to-electrolyte potentials manually and record the values and the time of each reading. Plot the results to see any trends.

5.4.9 If the test section is less than 1.6 km (1.0 mi), a single data logger may be installed at the test site.

5.4.10 Record each test station structure-to-electrolyte potential with another data logger for a period of 5 min.

5.5 Investigate the Cause of Poor Cathodic Protection

5.5.1 When practical, identify the source of any problems encountered during the commissioning survey.[9]

5.5.2 Inspect the DC power sources and compare with the target DC volts and DC amperes. If the target outputs are significantly different from the target, complete the following inspections:

5.5.2.1 If there is 0 to ~ 2 V_{DC}[10] and 0 A_{DC} output, look for trouble in the rectifier or the AC supply to the rectifier. Confirm that the ~2 V_{DC} reading is from the DC power source by turning it off and disconnecting one DC cable. If the reading stays at ~2 V_{DC}, then it is from the DC power source. If it drops to 0 V, then it is the galvanic difference between the structure material and anode or the carbon in the coke breeze.

5.5.2.2 If there is a normal voltage output but 0 A_{DC} output, look for the trouble in the cables, anodes, or connections external to the rectifier.

5.5.2.3 If there is approximately one-half normal voltage and approximately one-half normal current, then investigate the possibility of a failed diode, causing the rectifier to half-wave.

5.5.2.4 If a problem is suspected in the rectifier or the external DC circuit, refer to "Cathodic Protection Procedure No. 1: Rectifier Adjustment, Inspection, and Basic

[9] *See* Cathodic Protection Procedure No. 4: Diagnostic Testing (Current Requirement).

[10] The 2 V_{DC} may be the galvanic difference between the steel in the structure and the carbon in the coke breeze.

Troubleshooting." Complete repairs before proceeding with further testing.

5.5.3 Inspect the DC bonds and repair any broken bonds found.

5.5.4 Test all isolating features.(11)

5.5.5 Test all road or railroad casings, if applicable, to confirm that they are isolated.(12)

5.5.6 If the problems causing the loss of CP when corrected restore protection, then the troubleshooting is completed.

5.5.7 When the preceding faults are corrected but a CP criterion has still not been met, proceed with a DT.

5.6 Pipeline Current Measurements

5.6.1 Where practical, measure the current in a pipeline at intervals to determine the distribution of current with the current both OFF and ON.(13)

5.6.2 If current spans exist, measure the millivolt drop and polarity across the wires of a two-wire span or the inside wires of a four-wire span (see Figure 5.5) with the current both ON and OFF.

5.6.3 Where the pipeline comes aboveground, measure the current in the pipeline with a clamp-on ammeter. This measurement should be taken with the current applied both ON and OFF where the pipeline exits and enters the ground to note the amount of current pickup on the structure in between.

5.7 Alternating Current Voltage on Structures

5.7.1 Measure the AC structure-to-ground voltage before taking a DC structure-to-electrolyte potential to confirm that a hazardous voltage does not exist.

5.7.2 If an AC structure-to-ground voltage of 15 V_{AC} or greater exists, then follow the guidelines in NACE SP0177[7.1] and advise other personnel that the conditions exist.

---

11 *See* Cathodic Protection Procedure No. 9: Electrical Isolation.

12 *See* Cathodic Protection Procedure No. 10: Road and Railroad Cased Crossings (Basic).

13 *See* Cathodic Protection Procedure No. 3: Direct Current Measurements.

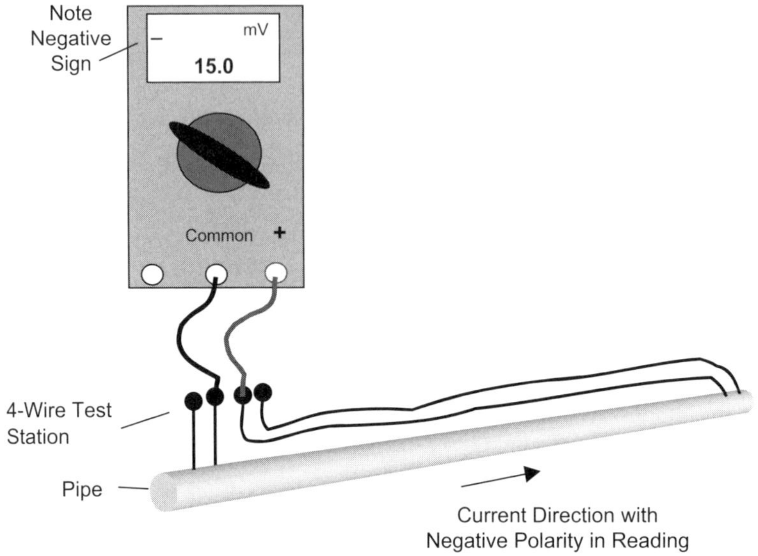

Figure 5.5 Current span measurement to calculate pipeline current is shown.

6.0 ANALYSIS

6.1 Criteria to Be Met

6.1.1 CP criteria are detailed in Section 6 of NACE SP0169.[7.2] Similar criteria are given in CGA OCC-1[7.3] and ISO 15589-1.[7.4 (14)] Test procedures to determine these criteria are given in NACE TM0497.[7.5]

6.1.2 There are three structure-to-electrolyte potential criteria for submerged or buried steel structures in the absence of specific data that demonstrate that adequate CP has been applied, including the following:

6.1.2.1 A negative (cathodic) potential of at least 850 mV (with respect to a CSE contacting the electrolyte) with the cathodic protection applied but with voltage drops other than across the structure to electrolyte interface (*IR* drop) removed. *IR* drops between the reference electrode and the structure-to-electrolyte boundary are an error in this

[14] Use most recent version. Note ISO 15589-1 includes only a polarized potential and polarization criteria.

reading and must be removed from the ON potential before applying to this criterion, as illustrated by Equation (6.1). NACE SP0169[7.2] discusses methods to evaluate the *IR* drop, which is essentially the difference between the ON potential and the instant OFF potential:

$$E_c = (E_{\text{on}} - IR), \tag{6.1}$$

where

- E_c potential for criterion ($-850\ \text{mV}_{\text{CSE}}$ or more negative)
- E_{on} potential with current applied (millivolts)
- IR voltage drop between the reference electrode and the structure-to-electrolyte boundary (millivolts)

6.1.2.2 A negative polarized potential equal to or more electronegative than -850 mV relative to a saturated copper-copper sulfate reference electrode contacting the electrolyte. This can be obtained by interrupting all current sources influencing the structure and taking an instant OFF potential.

6.1.2.3 A minimum of 100 mV of cathodic polarization between the structure surface and a stable reference electrode contacting the electrolyte. Polarization is the change in potential from the native or free corroding potential and the instant OFF potential, as illustrated by Equation (6.2), as determined by formation. Depolarization can also be determined by decay after all influencing power sources have been turned off (Equation [6.3]):

- Polarization

$$\Delta E_p = (E_{\text{off}} - E_{\text{native}}) \tag{6.2}$$

- Depolarization

$$\Delta E_{\text{depol}} = (E_{\text{off}} - E_{\text{depol}}), \tag{6.3}$$

where

ΔE_p	polarization for criterion (100-mV minimum)
E_{off}	potential with all current momentarily interrupted (millivolts)
E_{native}	native (free corroding) potential before CP current is applied (millivolts)
ΔE_{depol}	depolarization for criterion (100-mV minimum)
E_{depol}	depolarized potential with current left off (millivolts)

6.1.3 Only one of these criteria needs to be met. For example, if a polarized potential is less negative than −850 mV_{CSE}, a depolarization survey may prove that the 100-mV criterion is being achieved.

6.2 Cause of Subcriterion Potentials

6.2.1 Some common causes of a system's inability to meet criteria, tests, and cures are listed in Table 6.1.

6.2.2 Compare the DC power output to the historical information. If they are similar, review additional data.

6.3 Structure-to-Electrolyte Potentials

6.3.1 Confirm that all structure-to-electrolyte potentials meet one of the criteria for CP. Where readings do not meet this criterion, complete the subsequent analysis.

6.3.2 Readings that are more electronegative than during design tests at the same current output may suggest that the structure is exposed to anodic interference, it has been reduced in size, possibly by a faulty bond isolating part of the structure, or a CP system that is at a more electronegative potential is now shorted to the structure.

6.3.3 Readings that are more electronegative than before at the same current output may also suggest anodic interference; that is, a

Table 6.1 Summary of Identification of CP Trouble, Tests, and Causes

System Component	DC Power Source Volts	DC Power Source Amperes	Secondary AC Voltage	Structure-to-Electrolyte Potentials	Trouble Suspected	Tests	Remedy
Sacrificial Anode							
		A−		P−	May not be trouble. Anode current will decrease when potential becomes more negative.	No action	
		A+		P+	Opposite to above	Test for shorts or faults in the structure system.	Repair as required.
		A−		P+	Anodes failing	Measure anode-to-electrolyte potential. Perform anode voltage gradient test.	Replace anodes as necessary.
DC Power Source							
	0 to ~2*	0	0	P+	No AC power or failed DC power components. Trouble is before or in DC power source.	Confirm AC supply, check circuit breaker, test fuses, poor connections, or broken wires in DC power source. Check for signs of heat. If a battery supplemented source, test batteries for charge.	Confirm cause and correct before re-energizing
						If circuit breaker trips, look for short.	If no short, reduce DC output voltage.

(continued)

Table 6.1 *(continued)*

System Component	DC Power Source Volts	DC Power Source Amperes	Secondary AC Voltage	Structure-to-Electrolyte Potentials	Trouble Suspected	Tests	Remedy
	0 to ~2*	0	0	P+	No secondary AC voltage at taps	Test AC supply and circuit breaker.	If AC supply and circuit breaker are OK, then test transformer.
	0 to ~2*	0	V	P+	Fuse(s) in DC power source, failed rectifying element, poor connections, or faulty wire.	Test fuse(s), rectifying element, connections, and wires.	If fuse, test for short. If none found, lower voltage and re-energize. Otherwise, replace or repair as required.
	~ 1/2 V	~ 1/2 A	V	P+	One-half wave DC output. One part of rectifier bridge circuit is open.	Turn off, remove rectifying element connections, and test each diode or element.	Replace rectifying element(s).
	?V	?A	V	P	Faulty meters	Calibrate meters.	Replace as necessary.
DC Cables/Anode Bed							
	V	0	V	P+	Faulty cable, connections, or anodes	Trace cable to structure and anodes. Perform anode voltage gradient test.	Repair or replace as required.

	V	Dropping over time	V	P+	Failing or dry anodes	Potential profile over anodes to confirm status. Turn off to note recovery.	Temporary cure may be to water anodes. Replace anodes as required.
Structure							
	V	A	V	P+	Shorted isolation, accidental contact to foreign structure, faulty bonds, deteriorating coating	Test isolation and bonds. Trace for contacts. Complete DT test.	Repair isolation or bonds. Separate any contacts to foreign structures. Add CP capacity for poor coating or recoat.

Legend:
V Normal voltage
A Normal current
P Normal structure-to-electrolyte potential
V+ Greater than normal voltage
V− Lower than normal voltage
A+ Greater than normal current
A− Lower than normal current
?V, ?A Abnormal readings or varying
P+ Structure-to-electrolyte potential more electropositive
P− Structure-to-electrolyte potential more electronegative
* ~2 V may be due to galvanic difference between steel and anode or carbon in the coke breeze and not an indication of power

ground or anode bed of a foreign DC power source is in close proximity to the structure.

6.3.4 Readings that are less electronegative than during design tests may suggest a section of poor coating, poor electrical continuity, shorted isolation, a contact to a foreign structure, or shorted casings.

6.4 Dynamic Stray Current

6.4.1 If telluric or other dynamic stray currents are detected, the survey should be postponed until a quieter time, or if this is not practical, then data loggers should be installed at each end of the section to be tested to record pipe-to-electrolyte (pipe-to-soil) potentials versus time and should be left recording for a period of 22–24 hours.

6.4.2 If the test section is less than 1.6 km (1.0 mi), a single data logger may be installed at the test site.

6.4.3 Each test station pipe-to-electrolyte potential is to be recorded with another data logger for a period of 5 min. CIS potentials are taken as normal.

6.4.4 Telluric or other dynamic stray current activity requiring calibration is defined as an OFF potential fluctuation exceeding 20 mV peak to peak over the duration of the testing.

6.4.5 The true potential at the stationary data loggers must first be determined either by a quiet period or by an average over the test period. For each potential measurement along the line, the difference in potential between that and the stationary data logger at the same moment in time is to be determined. This difference added to the difference to the true potential at the stationary data logger is the correction factor for the portable data logger reading. Equations (6.4)–(6.6) are presented for the following situations:

- If two stationary data loggers are used, the method given in Equations (6.4) and (6.5) applies:

$$\acute{\varepsilon}_b = [\acute{\varepsilon}_a(c - b)/c] + [\acute{\varepsilon}_c(b - a)/c], \tag{6.4}$$

where

a	first stationary potential location
b	portable potential location
c	second stationary potential location
$\acute{\varepsilon}_a$	error in potential at a at time x
$\acute{\varepsilon}_b$	error in potential at b at time x
$\acute{\varepsilon}_c$	error in potential at c at time x

and

$$E_p = E_{p\ \text{measured}} - \acute{\varepsilon}_b, \tag{6.5}$$

where

E_p	true potential at the portable data logger location
$E_{p\ \text{measured}}$	potential at the portable data logger location

- If only one stationary data logger is used, then Equation (6.6) applies:

$$E_p = E_s - (E_{sa} - E_{pa}), \tag{6.6}$$

where

E_p	true potential at the portable data logger location
E_s	true potential at the stationary location
E_{sa}	stationary potential at time a during the data logging
E_{pa}	portable potential at time a during the data logging

6.4.6 Other methods can also be used to correct for dynamic stray currents.

6.4.7 The stationary recorder will also serve to verify synchronized interruption and if depolarization occurred during the survey interruption period.

6.5 Direct Current Power Source Interruption

6.5.1 Install a stationary data logger to monitor structure-to-electrolyte potentials. Review stationary data logger profiles each day to confirm that the interrupters at all influencing rectifiers continued to function during the test.

6.5.2 Identify any structure-to-electrolyte potential readings that may have been affected by an interrupter malfunction.

6.5.3 Note the amount of depolarization that took place during the interruption period.

6.6 Pipeline Current Measurement

6.6.1 Pipeline Current Span Method

6.6.1.1 If the resistance of the pipeline current span is known, calculate the ON and OFF current in the pipeline at each location, measured by Equation (6.7):

$$I_{span} = \frac{V_{span}}{R_{span}}, \tag{6.7}$$

where

I_{span} current in the pipeline current span (amperes)
V_{span} voltage drop across the current span (volts)
R_{span} resistance of the current span (ohms)

6.6.1.2 If the calibration of the pipeline current span is known, calculate the ON and OFF current in the pipeline at each location, measured by Equation (6.8):

$$I_{span} = \mathrm{mV}_{span} \times \mathrm{CF}_{span}, \tag{6.8}$$

where

I_{span} current in the pipeline current span (amperes)
mV_{span} voltage drop across the current span (millivolts)
CF_{span} calibration factor of the current span (amperes per millivolt)

6.6.2 Clamp-On Ammeter

Calculate an average of the current measured in both directions and note the actual direction of current.

6.6.3 Current Pickup

Calculate the current pickup between pipeline current measurement locations. The current will normally be in a direction pointing toward the nearest current source and increasing as it approaches.

7.0 REFERENCES

7.1 NACE Standard SP0177-2007, "Mitigation of Alternating Current and Lightning Effects on Metallic Structures and Corrosion Control Systems" (Houston, TX: NACE International, 2007).

7.2 NACE Standard SP0169-2007, "Control of External Corrosion on Underground or Submerged Metallic Piping Systems" (Houston, TX: NACE International, 2007).

7.3 CGA Recommended Practice OCC-1-2005, "Control of External Corrosion on Buried or Submerged Metallic Piping Systems" (Canadian Gas Association, Ottawa, Ontario, 2005).

7.4 ISO 15589-1, "Petroleum and Natural Gas Industries: Cathodic Protection of Pipeline Transportation Systems, Part 1—On-Land Pipelines" (Geneva, Switzerland: ISO, 2003).

7.5 NACE Standard TM0497-2002, "Measurement Techniques Related to Criteria for Cathodic Protection on Underground or Submerged Metallic Piping Systems" (Houston, TX: NACE International, 2002).

7.6 A.W. Peabody, *Control of Pipeline Corrosion*, 2nd ed., ed. R.L. Bianchetti (Houston, TX: NACE International, 2001), p. 65.

CATHODIC PROTECTION PROCEDURE NO. 7

Close Interval Potential Survey 7

1.0 INTRODUCTION

The purpose of a close interval structure-to-electrolyte potential survey (CIS) procedure is to outline a process that will obtain a more representative series of potentials throughout the structure to evaluate the criteria for cathodic protection, to identify local deficiencies in a cathodic protection system, to assist in identifying the cause(s) of such deficiencies, and to aid in determining the requirements to enhance the level of cathodic protection to the structure.

The CIS survey obtains structure-to-electrolyte potentials at relatively short intervals along the structure such as at 1-m (3-ft) to 3-m (10-ft) spacing.

2.0 TOOLS AND EQUIPMENT

The following equipment will vary depending on the test selected:

- Multimeter capable of measuring 1 mV_{DC} to 40 V_{DC}, complete with leads with insulated probes
- Portable direct current (DC) voltmeter data logger(s)
- Copper-copper sulfate reference electrode(s), complete with extension
- Multimeter including alternating current (AC)/DC volts and an ohmmeter

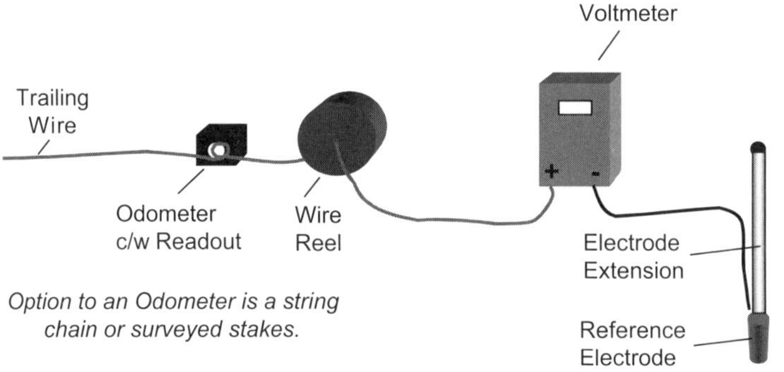

Figure 2.1 Basic components of a CIS with a voltmeter are shown.

- Synchronized current interrupter(s), preferably Global Positioning System (GPS)-synchronized
- Wire reel and odometer
- Battery and control resistor or portable controlled DC power supply
- Pipe locator transmitter and receiver
- Test wires as necessary for the applicable test
- Small hand tools

A typical connection for a CIS using manually recorded data is shown in Figure 2.1.

A typical connection for a CIS with data stored in a memory as it is collected is shown in Figure 2.2.

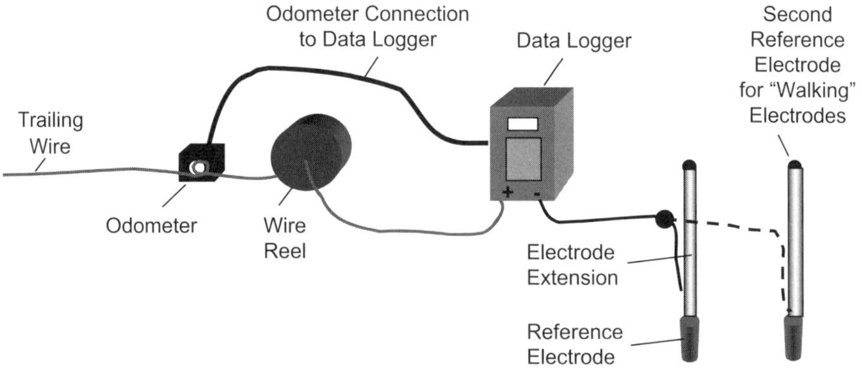

Figure 2.2 Typical components for a CIS using a data logger are shown.

The second reference electrode in Figure 2.2 is required for the walking reference electrode survey method. The two reference electrodes are connected together in this method.

3.0 SAFETY EQUIPMENT

- Standard safety equipment and clothing, as required by the company's safety manual and regulations
- Electrically insulated clips and probe handles for meter leads
- Only personnel who have received training and are qualified in accordance with local codes and regulations are to work on DC power sources or their supply.
- A hazardous AC voltage can be induced on a long conductor that parallels an AC power line.[7.1] Special shielding may be necessary in these conditions.
- When working in traffic, arrange to have the necessary traffic control.

4.0 PRECAUTIONS

The following precautions are in addition to those that must also be followed when working on a particular facility.

4.1 Measure the voltage between the rectifier case and ground before touching the case.

4.2 Open the case expecting to find biting insects, rodents, or snakes inside and take appropriate precautions.

4.3 Inspect the rectifier for abnormal sounds, temperature, or odors and, if noted, turn it off.

4.4 Switch off the AC voltage supply before installing a current interrupter or each time the taps are adjusted.

4.5 Secure any exposed electrical terminals in a locked container when the rectifier is not attended.

4.6 Measure a structure AC voltage to ground on the structure before taking cathodic protection measurements. If the AC voltage to ground is equal to or exceeds 15 V_{AC}, practice the safety measures detailed in NACE SP0177[7.1] and advise other personnel working on the structure of the hazard.

4.7 When working near high-voltage AC (HVAC) power lines, take AC structure-to-ground voltage readings at frequent intervals as these voltages can change with the power line load, geometry, and the length of the trailing wire.[7.1]

4.8 Do not work on the structure when lightning is in the area.

4.9 When working near a fence, confirm that it is not an electric fence for livestock (look for insulators) and that an AC voltage is not being induced on it by a parallel HVAC power line.

4.10 Run the trailing wire under any fences or in such a manner that it does not come in contact with other metal structures.

5.0 PROCEDURE

There are two basic methods of using the reference electrode, that is, either by a single reference electrode or by two walking reference electrodes. The data can in turn be collected manually or saved in a memory of a data logger. Variations of these approaches can be employed, depending on the circumstances.

5.1 Voltmeter with Single Reference Electrode Measuring ON Potentials

5.1.1 Using a pipe locator, stake the route of the pipeline to be surveyed. If an odometer is not used, then the stakes should reflect the distance interval along the pipe to be recorded during the CIS.

5.1.2 Connect the wire from the wire reel between the first test station and the voltmeter's positive terminal (see Figure 5.1).

5.1.3 Connect the negative terminal to the copper-copper sulfate reference electrode (CSE). Another type of reference electrode can be substituted, depending on the electrolyte.

5.1.4 Measure the structure-to-electrolyte potential.[(1)]

5.1.5 Measure a structure-to-electrolyte potential to the same CSE position, but connecting the voltmeter positive directly to the test station, rather than through the wire in the reel.

---

[1] *See* Cathodic Protection Procedure No. 2: Structure-to-Electrolyte Potential Measurement.

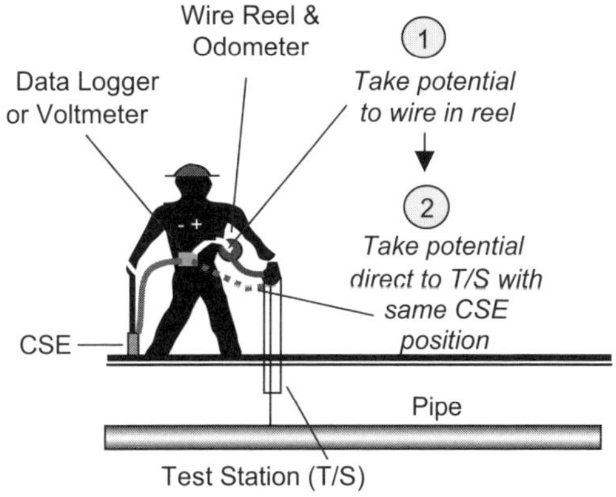

Figure 5.1 Procedure at the start of CIS is shown.

5.1.6 Confirm that the potential measurements obtained following the procedures in Sections 5.1.4 and 5.1.5 are equal; if they are not, check the connections. Do not proceed until these readings are the same.

5.1.7 Secure the wire from the reel to the test station, making certain that it does not contact any metal parts.

5.1.8 Set up the odometer to read zero at the test station and make certain that it will advance in accordance with the forward motion.

5.1.9 Set up the field notes to record the date, time, and tester, with distance and potential columns.

5.1.10 Move forward to the next reading location and place the CSE into moist soil.

5.1.11 After making certain that the next structure-to-electrolyte potential is stable and accurate, record the distance and potential, including polarity, value, and units (Figure 5.2).

5.1.12 Proceed to the next reading and repeat the procedure detailed in Section 5.1.10.

5.1.13 When the next test station is reached, measure the structure-to-electrolyte potential to the wire in the reel (test station 1 in Figure 5.3).

5.1.14 Connect a wire directly from the next test station to the voltmeter (test station 2 in Figure 5.3) and measure the structure-to-electrolyte potential.

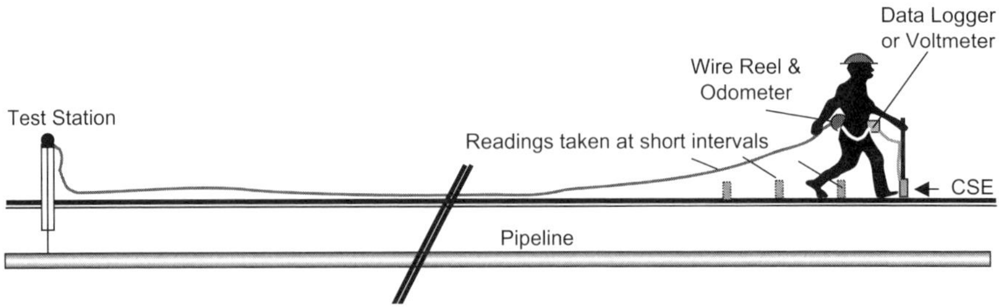

Figure 5.2 Single reference electrode CSI procedure is shown.

5.1.15 In this case (Figure 5.3), the potential taken to test station 1 can be different from that taken to test station 2 due to the *IR* drop from the current in the pipeline. Depending on the direction of current, this *IR* drop will either add to or subtract from the potential taken to the last test station (test station 1 in Figure 5.3).

5.1.16 If the difference between the potential at test stations 1 and 2 is quite different, a correction should be applied, as given in Equation (6.1) or (6.2) in Section 6.0.

5.1.17 Continue the survey by repeating the procedures in Sections 5.1.1–5.1.15.

5.2 Data Logger with Single Reference Electrode Measuring ON Potentials

5.2.1 Using a pipe locator, stake the route of the pipeline to be surveyed. If an odometer is not used, then the stakes should reflect the distance along the pipe to be recorded during the CIS.

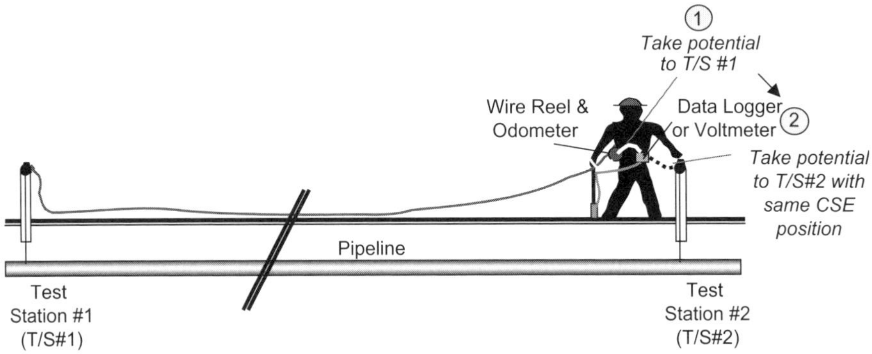

Figure 5.3 Procedure at next test station in a CIS is shown.

5.2.2 Connect the wire from the wire reel between the first test station and the data logger's positive terminal (see Figure 5.1).

5.2.3 Connect the negative terminal to the CSE. Another reference electrode can be substituted, depending on the electrolyte.

5.2.4 Measure the structure-to-electrolyte potential.[(2)]

5.2.5 Measure a structure-to-electrolyte potential to the same CSE, but connect the data logger's positive terminal directly to the test station, rather than through the wire in the reel.

5.2.6 Confirm that the two potential measurements obtained following the procedures in Sections 5.2.4 and 5.2.5 are equal; if they are not, check the connections. If the data logger is equipped with multiple input impedances, select a higher impedance to see if the potentials are closer. If so, a high resistance exists in the circuit either in the connections or in the reference electrode's contact to earth. Do not proceed until these readings are the same.

5.2.7 Secure the wire from the reel to the test station, making certain that it does not contact any metal parts.

5.2.8 Set up the odometer to read zero at the test station and make certain that it will advance in accordance with the forward motion.

5.2.9 Connect the odometer cable to the data logger.

5.2.10 Record the segment to be tested, date, time, and tester in the data logger.

5.2.11 Enter the first structure-to-electrolyte potential.

5.2.12 Move forward to the next reading location and place the CSE into moist soil.

5.2.13 When the reading is stable and accurate, enter the next reading into the data logger.

5.2.14 When the next test station is reached, measure the structure-to-electrolyte potential to the wire in the reel (test station 1 in Figure 5.3).

5.2.15 Connect a wire directly from the next test station to the data logger (test station 2 in Figure 5.3) and measure the structure-to-electrolyte potential; record and label it accordingly.

---

[2] *See* Cathodic Protection Procedure No. 2: Structure-to-Electrolyte Potential Measurement.

5.2.16 In Figure 5.3, the potential taken to test station 1 can be different from that taken to test station 2 due to the *IR* drop from the current in the pipeline. Depending on the direction of current, this *IR* drop will either add to or subtract from the potential taken to the last test station (test station 1 in Figure 5.3).

5.2.17 If the difference between the potential at test stations 1 and 2 is quite different, a correction should be applied, as given in Equation (6.1). This correction assumes that the pipeline *IR* drop is uniform in this short distance.

5.2.18 Download the data from the data logger regularly.

5.3 Data Logger with Walking Reference Electrode Measuring ON Potentials

5.3.1 Using a pipe locator, stake the route of the pipeline to be surveyed. If an odometer is not used, then the stakes should reflect the distance along the pipe to be recorded during the CIS.

5.3.2 Connect the wire from the wire reel between the first test station and the data logger's positive terminal (see Figure 5.1).

5.3.3 Connect the negative terminal to the CSE. Another reference electrode can be substituted, depending on the electrolyte.

5.3.4 Measure the structure-to-electrolyte potential.[(3)]

5.3.5 Measure a structure-to-electrolyte potential to the same CSE, but connect the data logger's positive terminal directly to the test station, rather than through the wire in the reel.

5.3.6 Confirm that the two potential measurements obtained following the procedures in Sections 5.3.4 and 5.3.5 are equal; if they are not, check the connections. If the data logger is equipped with multiple input impedances, select a higher impedance to see if the potentials are closer. If so, a high resistance exists in the circuit, either in the connections or in the reference electrode's contact to earth. Do not proceed until these readings are the same.

---

[3] *See* Cathodic Protection Procedure No. 2: Structure-to-Electrolyte Potential Measurement.

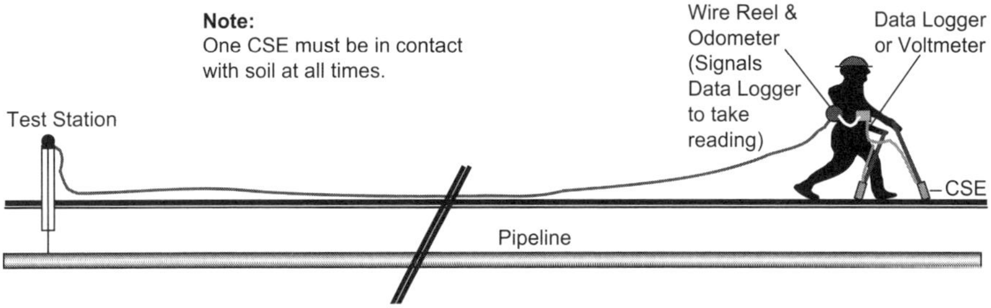

Figure 5.4 CIS using a walking reference electrode is shown.

5.3.7 Secure the wire from the reel to the test station, making certain that it does not contact any metal parts.

5.3.8 Make certain that the odometer will advance in accordance with the forward motion and set it to read zero at the test station.

5.3.9 Connect the odometer cable to the data logger.

5.3.10 Record the segment to be tested, date, time, and tester in the data logger.

5.3.11 Set the data logger to record readings at the intervals required.

5.3.12 Enter the first structure-to-electrolyte potential.

5.3.13 Move forward, making certain that one CSE is in good contact with the soil at all times as the tester is not certain exactly when the data logger will capture the reading (see Figure 5.4).

5.3.14 When the next test station is reached, measure the structure-to-electrolyte potential to the wire in the reel (test station 1 in Figure 5.3).

5.3.15 Connect a wire directly from the next test station to the data logger (test station 2 in Figure 5.3) and measure the structure-to-electrolyte potential.

5.3.16 In Figure 5.3, the potential taken to test station 1 can be different from that taken to test station 2 due to the *IR* drop from the current in the pipeline. Depending on the direction of current, this *IR* drop will either add to or subtract from the potential taken to the last test station (test station 1 in Figure 5.3).

5.3.17 If potentials at test stations 1 and 2 are quite different, a correction should be applied, as given in Equation (6.1). This correction assumes that the pipeline *IR* drop is uniform with distance.

5.3.18 Download the data from the data logger regularly.

5.4 Voltmeter with Single Reference Electrode Measuring ON/OFF Potentials

5.4.1 Using a pipe locator, stake the route of the pipeline to be surveyed. If an odometer is not used, then the stakes should reflect the distance along the pipe to be recorded during the CIS.

5.4.2 Install a synchronized current interrupter in all DC power sources and bonds and record the time for the ON and OFF cycles. Preferably, set the cycles so that they divide evenly into a minute and start the cycles at the even minute so that the tester can confirm which cycle exists, using a watch.

5.4.3 Connect the wire from the wire reel between the first test station and the voltmeter's positive terminal (see Figure 5.1).

5.4.4 Connect the negative terminal to the CSE. Another type of reference electrode can be substituted, depending on the electrolyte.

5.4.5 Set up the field notes to record the date, time, and tester, with distance and potential columns.

5.4.6 Measure the ON and OFF structure-to-electrolyte potentials[(4)] and monitor them until they are stable. If using a digital voltmeter, take the second clear reading after the DC power sources go off as the instant OFF potential. If the potentials are not stable, follow the procedure given in Section 5.7.

5.4.7 Measure the ON and OFF structure-to-electrolyte potential to the same CSE, but connecting the voltmeter positive directly to the test station, rather than through the wire in the reel.

5.4.8 Confirm that the two ON and the two OFF potential measurements obtained following the procedures in Sections 5.4.6 and 5.4.7 are equal; if they are not, check the connections. Do not proceed until these readings are the same.

5.4.9 Secure the wire from the reel to the test station, making certain that it does not contact any metal parts.

---

[4] *See* Cathodic Protection Procedure No. 2: Structure-to-Electrolyte Potential Measurement.

5.4.10 Set up the odometer to read zero at the test station and make certain that it will advance in accordance with the forward motion.

5.4.11 Move forward to the next reading location and place the CSE into moist soil.

5.4.12 Identify which reading is the ON potential and which is the OFF potential. Do not assume that the more negative potential is the ON potential. Identify the ON/OFF potentials by the length of the cycles. Using a watch, if the cycles were started at the even minute and if they divide evenly into a minute, can distinguish the cycle.

5.4.13 After making certain that the next ON/OFF structure-to-electrolyte potentials are stable and accurate, record the distance and the ON/OFF potentials, including polarity, value, and units (Figure 5.2).

5.4.14 Proceed to the next reading and repeat the procedure detailed in Section 5.1.11.

5.4.15 When the next test station is reached, measure the ON/OFF structure-to-electrolyte potentials to the wire in the reel (test station 1 in Figure 5.3).

5.4.16 Connect a wire directly from the next test station to the voltmeter (test station 2 in Figure 5.3) and measure the ON/OFF structure-to-electrolyte potentials again.

5.4.17 These values can be different due to the *IR* drop in the pipeline. Equation (6.1) can make corrections for this *IR* drop, assuming that the *IR* drop is linear.

5.5 Data Logger with Single Reference Electrode Measuring ON/OFF Potentials

5.5.1 Using a pipe locator, stake the route of the pipeline to be surveyed. If an odometer is not used, then the stakes should reflect the distance along the pipe to be recorded during the CIS.

5.5.2 Install a synchronized current interrupter in all DC power sources and bonds and record the time for the ON and OFF cycles. Preferably, set the cycles so that they divide evenly into a minute and start the cycles at the even minute so that the tester can confirm which cycle exists using a watch.

5.5.3 Install a stationary data logger to monitor the interruption cycles and dynamic stray current (see Section 5.7).

5.5.4 Connect the wire from the wire reel between the first test station and the data logger's positive terminal (see Figure 5.1).

5.5.5 Connect the negative terminal to the CSE. Another type of reference electrode can be substituted, depending on the electrolyte.

5.5.6 Determine which reading occurs during the ON cycle versus the OFF cycle.

5.5.7 Measure the ON/OFF structure-to-electrolyte potential.[(5)]

5.5.8 Measure an ON/OFF structure-to-electrolyte potential to the same CSE, but connect the data logger's positive terminal directly to the test station, rather than through the wire in the reel.

5.5.9 Confirm that the two ON and the two OFF potential measurements obtained following the procedures in Sections 5.5.7 and 5.5.8 are equal; if they are not, check the connections, or if the data logger is equipped with multiple input impedances, select a higher impedance to see if the potentials are closer. If so, a high resistance exists in the circuit, either in the connections or in the reference electrode's contact to earth. Do not proceed until these readings are the same.

5.5.10 Secure the wire from the reel to the test station, making certain that it does not contact any metal parts.

5.5.11 Set up the odometer to read zero at the test station and make certain that it will advance in accordance with the forward motion.

5.5.12 Connect the odometer cable to the data logger.

5.5.13 Record the location, segment to be tested, date, time, and tester in the data logger.

5.5.14 Determine the order of entering the ON versus the OFF structure-to-electrolyte potentials and maintain this order for all subsequent readings; that is, if the ON potential is entered first at the start, then it is to be entered first at all other reading

---

[5] *See* Cathodic Protection Procedure No. 2: Structure-to-Electrolyte Potential Measurement.

locations. This will save entering an "alpha" label for each reading.

5.5.15 Enter the first ON and OFF structure-to-electrolyte potentials.

5.5.16 Move forward to the next reading location and place the CSE into moist soil.

5.5.17 After the reading is stable and accurate, enter the next ON/OFF potential readings into the data logger.

5.5.18 When the next test station is reached, measure the ON/OFF structure-to-electrolyte potentials to the wire in the reel (test station 1 in Figure 5.3).

5.5.19 Connect a wire directly from the next test station to the data logger (test station 2 in Figure 5.3) and measure the ON/OFF structure-to-electrolyte potentials again.

5.5.20 In Figure 5.3, the potential taken to test station 1 can be different from that taken to test station 2 due to the *IR* drop from the current in the pipeline. Depending on the direction of current, this *IR* drop will either add to or subtract from the potential taken to the last test station (test station 1 in Figure 5.3).

5.5.21 If the difference between either the two ON potentials or the two OFF potentials taken to test stations 1 and 2 is quite different, a correction should be applied, as given in Equation (6.1). This correction assumes that the pipeline *IR* drop is uniform with distance.

5.5.22 With the negative of the data logger still connected to the CSE, connect the wire from the wire reel to the next test station and the data logger's positive terminal (see Figure 5.1) and repeat the procedure detailed in this section.

5.5.23 Download the data from the data logger and save regularly.

5.6 Data Logger with Walking Reference Electrode Measuring ON/OFF Potentials

5.6.1 Using a pipe locator, stake the route of the pipeline to be surveyed. If an odometer is not used, then the stakes should reflect the distance along the pipe to be recorded during the CIS.

5.6.2 Connect the wire from the wire reel between the first test station and the data logger's positive terminal (see Figure 5.1) and connect the negative terminal to the CSE. Another reference electrode can be substituted, depending on the electrolyte.

5.6.3 Install a stationary data logger to monitor the interruption cycles and dynamic stray current (see Section 5.7).

5.6.4 Record the location, test segment, date, time of day, cycle times, and tester in the data logger.

5.6.5 Determine which reading occurs during the ON versus the OFF interruption cycle and record the ON and OFF structure-to-electrolyte potentials.[(6)]

5.6.6 Measure a structure-to-electrolyte potential to the same CSE, but connect the data logger's positive terminal directly to the test station, rather than through the wire in the reel.

5.6.7 Confirm that the two potential measurements obtained following the procedures given in Sections 5.6.5 and 5.6.6 are equal; if they are not, check the connections. If the data logger is equipped with multiple input impedances, select a higher impedance to see if the potentials are closer. If so, a high resistance exists in the circuit, either in the connections or in the reference electrode's contact to earth. Do not proceed until these readings are the same.

5.6.8 Secure the wire from the reel to the test station, making certain that it does not contact any metal parts.

5.6.9 Make certain that the odometer will advance in accordance with the forward motion and set it to read zero at the first test station.

5.6.10 Connect the odometer cable to the data logger.

5.6.11 Set the data logger to record readings at the interval required.

5.6.12 Enter the first ON/OFF structure-to-electrolyte potentials. Instant OFF potentials are to be recorded at least 0.6 s after interruption to avoid recording a potential spike.

5.6.13 Move forward making certain that one CSE is in good contact with the soil at all times as the tester is not certain exactly when the data logger will capture the reading (Figure 5.4).

5.6.14 When the next test station is reached, measure the ON/OFF structure-to-electrolyte potentials to the wire in the reel (test station 1 in Figure 5.3).

---

[6] *See* Cathodic Protection Procedure No. 2: Structure-to-Electrolyte Potential Measurement.

5.6.15 Connect a wire directly from the next test station to the data logger (test station 2 in Figure 5.3) and measure the ON/OFF structure-to-electrolyte potentials again.

5.6.16 In Figure 5.3, the ON/OFF potential taken to test station 1 can be different from those taken to test station 2 due to the *IR* drop from the current in the pipeline. Depending on the direction of current, this *IR* drop will either add to or subtract from the potential taken to the last test station (test station 1 in Figure 5.3).

5.6.17 If the difference between either the two ON potentials or the two OFF potentials taken to test stations 1 and 2 is quite different, a correction should be applied, as given in Equation (6.1). This correction assumes that the pipeline *IR* drop is uniform with distance.

5.6.18 With the negative of the data logger still connected to the CSE, connect the wire from the wire reel to the next test station and the data logger's positive terminal (Figure 5.1) and repeat the work detailed in this section.

5.6.19 Download the data from the data logger and save regularly.

5.7 Interruption Cycles

5.7.1 When the interruption cycles deviate from the actual timing, or when potential levels suddenly increase or decrease suddenly, stop the CIS and determine the cause.

5.7.2 The possible causes of interruption timing failure include the following:

5.7.2.1 Interrupters falling out of synchronization with each other. This results in more than one ON or OFF potential during the same time frame, as shown in Figure 5.5. OFF cycle A in this figure is synchronized, and the true instant OFF potential can be read directly. OFF cycle B shows one interrupter going off just before the intended OFF cycle, and although the true OFF occurs for part of this time, there are two others to choose from. Cycle C shows an interrupter going off and on before the ON cycle is finished. In the last case, the instant OFF cannot be read

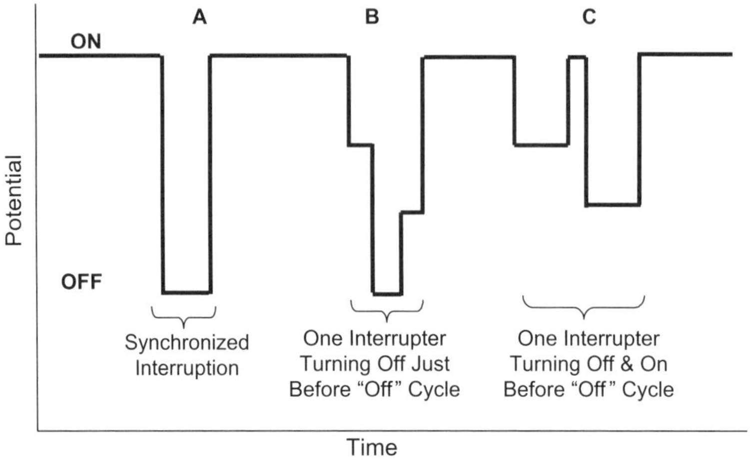

Figure 5.5 Effect of different current interruption synchronization on potentials is shown.

directly. The total *IR* drop can be calculated by taking the sum of the *IR* drops in the two OFF potentials. It is best to resynchronize the interrupters.

5.7.2.2 An interrupter failing in the ON position. This is typical of many GPS-synchronized interrupters when they lose the satellite signal. Unfortunately, the instant OFF potential is no longer a true polarized potential.

5.7.2.3 An interrupter fails in the OFF position, resulting in a loss of polarization.

5.7.3 Use a stationary data logger to identify any loss of interrupter synchronization and to salvage polarized potential data when this problem occurs.

5.8 Dynamic Stray Current

5.8.1 Install stationary data loggers during the CIS, especially if dynamic stray currents may be suspected.

5.8.2 Select a reading time cycle such that the memory is capable of storing the number of readings for the time that is to be data logged.

5.8.3 If the CIS section is approximately 1.0–3.0 km (0.6–1.8 mi), use a single stationary data logger. If the CIS is longer, use one data logger located near the one-fourth and three-fourths distance points of the CIS.

5.8.4 If telluric currents are being experienced, run the stationary data log through the survey and through the night as that is often when the quietest telluric activity period will occur. If the dynamic current is from other sources, run the data log during the survey and past a point where the stray current stops.

6.0 ANALYSIS

6.1 Voltmeter with Single Reference Electrode Measuring ON Potentials

6.1.1 Plot the data collected on a distance scale (x-axis) that will identify the individual readings taken and on a potential scale (y-axis) that will easily recognize the changes in value along the line.

6.1.2 Note that an ON potential without correction for *IR* drop cannot be used to confirm a criterion.[7.2,7.3] Areas that do not meet the criterion without the *IR* drop removed will, of course, not meet it afterward, and, thus, these areas can be identified as subcriterion, although it is likely that the limits will need to be expanded.

6.1.3 The ON potential profile will locate areas where bare metal or larger coating holidays are exposed to the soil.

6.1.4 If the structure-to-electrolyte potential taken to the next test station is significantly different than the one taken to the test station connected to the trailing wire, a correction should be applied. One correction, given in Equation (6.1), assumes that the pipeline *IR* drop is uniform with distance and that the potential at the next test station is more positive than that measured through the trailing wire to the starting test station. If the potential is more negative than that measured to the trailing wire, use Equation (6.2):

$$E_{\text{on}} = E_{\text{on test}} - \frac{(X - B)}{X} \times \Delta E_{1-2} \tag{6.1}$$

$$E_{\text{on}} = E_{\text{on test}} + \frac{(X - B)}{X} \times \Delta E_{1-2}, \tag{6.2}$$

where

E_{on}	ON corrected structure-to-electrolyte potential (volts)
$E_{on\ test}$	ON structure-to-electrolyte at distance A (volts)
X	total distance of survey segment (meters or feet)
A	distance from the start of the survey segment to reading location (meters or feet)
B	remaining distance from reading location to next test station ($X - A$) (meters or feet)
ΔE_{1-2}	difference in potential between the first and next test stations (volts)

6.2 Data Log with Single Reference Electrode Measuring ON Potentials

6.2.1 The data taken from the memory of the data logger can be plotted directly by a computer.

6.2.2 The analysis in Section 6.1 then applies.

6.3 Data Log with Walking Reference Electrode Measuring ON Potentials

6.3.1 The data downloaded from the memory of the data logger can be plotted directly by a computer.

6.3.2 The analysis in Section 6.1 then applies.

6.4 Voltmeter with Single Reference Electrode Measuring ON/OFF Potentials

6.4.1 Plot the ON and OFF structure-to-electrolyte potentials as separate profiles on a distance scale (x-axis) that will identify the individual readings taken and on a potential scale (y-axis) that will easily recognize the changes in potential along the line.

6.4.2 Note that the OFF (polarized) potential is used to confirm that the criterion for cathodic protection has been met.[7.2,7.3] In addition, plot a -850-mV_{CSE} line to help identify subcriterion sections of pipeline.

6.4.3 The ON potential profile will better locate areas where bare metal or larger coating holidays are exposed to the soil due to a decrease in the voltage (IR) drop across the coating coincident with the holiday or defect.

6.4.4 If the structure-to-electrolyte potential(s) taken to the next test station is significantly different than the one taken to the test station connected to the trailing wire, a correction should be applied. Calculate the ON potential from either Equation (6.1) or Equation (6.2) and calculate the OFF potential using Equation (6.3):

$$E_{\text{off}} = E_{\text{on}} - \Delta E_{\text{on-off}}, \tag{6.3}$$

where

E_{off}	corrected OFF structure-to-electrolyte potential at reading location (volts)
E_{on}	corrected ON structure-to-electrolyte potential at reading location from Equation (6.1) or (6.2; volts)
$\Delta E_{\text{on-off}}$	ON potential minus OFF potential at reading location (volts)

6.5 Data Log and Single Reference Electrode Measuring ON/OFF Potentials

6.5.1 The data downloaded from the memory of the data logger can be plotted directly by a computer; however, if drawn as a solid line, it will appear as a band between the ON and OFF readings because they alternate in the same column. An option is to plot them as data points or separate the ON and OFF potentials with another program.

6.5.2 The analysis in Section 6.4 then applies.

6.6 Data Log Using Walking Reference Electrode Measuring ON/OFF Potentials

6.6.1 Because the potentials from the walking reference electrode method are taken as the tester is moving, the ON and OFF potentials are not taken at exactly the same location. The data plotted directly will give a stepped ON and OFF profile versus distance along the pipeline, such as shown in Figure 6.1, as the ON and OFF are not taken at exactly the same location. A faster interruption cycle reduces this problem.

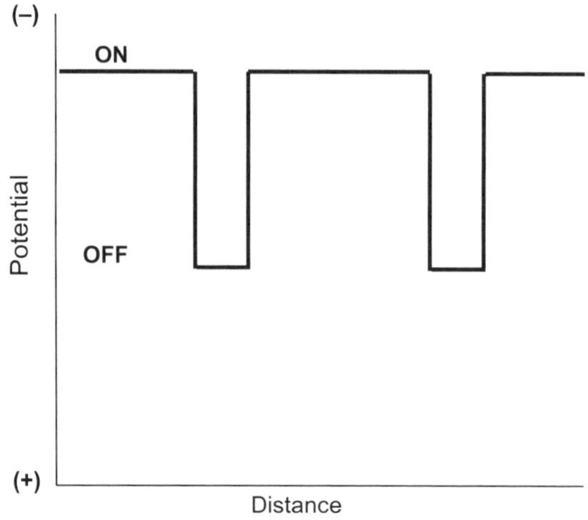

Figure 6.1 Stepped potentials over distance with walking reference electrode raw data are shown.

6.6.2 A program can plot separate ON and OFF potential profiles, thereby eliminating the stepped effect, provided that the ON and OFF potentials are identified correctly. Do not assume that the more negative potential is always the ON potential. It is wise to plot the stepped profile first to confirm the ON and OFF cycles by the time of the cycles, rather than by their values. This will also assist in the analysis if the interrupters go out of synchronization (Figure 5.5).

6.6.3 The analysis in Section 6.4 then applies.

6.7 Interruption Cycles

6.7.1 Where the interruption cycles are synchronized, then proceed with the appropriate preceding analysis (example A in Figure 6.2).

6.7.2 If there is a loss of synchronization that is less than the length of the shorter cycle, then the largest *IR* drop is then applied (example B in Figure 6.2). Assuming that the OFF potential is less negative than the ON potential, then the least negative value during the OFF cycle is the polarized potential.

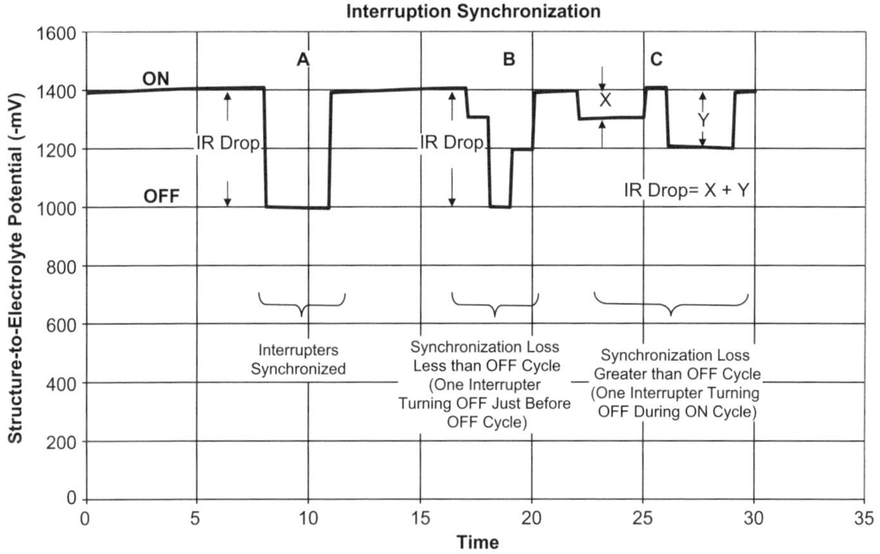

Figure 6.2 Loss of interrupter synchronization is shown.

6.7.3 If the loss of synchronization occurs before or after the normal cycle, and the *IR* drop values are clearly indicated, they can be summed together and the OFF potential calculated (example C in Figure 6.2). The OFF potential in example C in Figure 6.2 is then calculated from the ON potential using Equation (6.4):

$$E_{\text{off}} = E_{\text{on}} - X - Y, \tag{6.4}$$

where

E_{off} true OFF or polarized potential
E_{on} ON potential
X *IR* drop from the DC power source not synchronized
Y *IR* drop from the DC power source(s) synchronized

6.7.4 Where the loss of synchronization is such that there is no clear cycle from the DC power sources, the survey must be repeated.

6.8 Dynamic Stray Current

6.8.1 There are different methods to analyze dynamic current; one of these methods is given subsequently. This method requires that at least one stationary data logger was in place and that the stationary and portable data logger timing was synchronized.

6.8.2 Where two stationary data loggers are used, the potentials can be corrected using Equation (6.5):

$$e_b = [e_a(c - b)/c] + [e_c(b - a)/c], \tag{6.5}$$

where

- a first stationary potential location (meters or feet)
- b portable potential location (meters or feet)
- c second stationary potential location (meters or feet)
- e_a error in potential at a at time x (volts)
- e_b error in potential at b at time x (volts)
- e_c error in potential at c at time x (volts)

Equation (6.6) then applies the correction to the measured potential at the portable data logger location:

$$E_{p\ \text{true}} = E_{p\ \text{measured}} - \acute{\varepsilon}_b. \tag{6.6}$$

In this particular case, however, the potential closely follows that of the nearby stationary data logger position, and correcting only to the first stationary data log location would be accurate. Equation (6.7) is then used to compensate the measured potential:

$$E_p = E_s - (E_{sa} - E_{pa}), \tag{6.7}$$

where

- E_p true potential at the portable data logger location (volts)
- E_s true potential at the stationary location (volts)
- E_{sa} stationary potential at time a during the data logging (volts)
- E_{pa} portable potential at time a during the data logging (volts)

6.9 Criteria

6.9.1 Cathodic protection criteria are detailed in Section 6.0 of NACE SP0169.[7.2] Similar criteria are given in CGA OCC-1[7.3] and ISO 15589-1.[7.4] [(7)] Other test procedures to determine these criteria are given in NACE Standard TM0497.[7.5]

6.9.2 There are three structure-to-electrolyte potential criteria for submerged or buried steel structures in the absence of specific data that demonstrate that adequate cathodic protection has been applied, including the following:

- A negative (cathodic) potential of at least 850 mV with the cathodic protection applied (with respect to a CSE contacting the electrolyte) but with voltage drops other than across the structure-to-electrolyte interface (*IR* drop) removed. *IR* drops between the reference electrode and the structure-to-electrolyte boundary are an error in this reading and must be removed from the ON potential before applying to this criterion, as illustrated by Equation (6.8). NACE SP0169[7.2] discusses methods to evaluate the *IR* drop, which is essentially the difference between the ON potential and the instant OFF potential:

$$E_c = (E_{\text{on}} - IR), \tag{6.8}$$

where

E_c potential for criterion ($-850\ \text{mV}_{\text{CSE}}$ or more negative)

E_{on} potential with current applied (millivolts)

IR voltage drop between the reference electrode and the structure-to-electrolyte boundary (millivolts)

- A negative polarized potential equal to or more electronegative than −850 mV relative to a saturated copper-copper sulfate reference electrode contacting the electrolyte. This can be obtained by interrupting all current sources influencing the structure and taking an instant OFF potential.

---

[7] Use most recent version. Note ISO 15589-1 includes only a polarized potential and polarization criteria.

- A minimum of 100 mV of cathodic polarization between the structure surface and a stable reference electrode contacting the electrolyte. Polarization is the change in potential from the native or free corroding potential and the instant OFF potential, as illustrated by Equation (6.9), as determined by formation. Depolarization can also be determined by decay after turning all rectifiers off and conducting a separate depolarized potential survey (Equation [6.10]):
  - Polarization

$$\Delta E_p = (E_{\text{off}} - E_{\text{native}}) \tag{6.9}$$

  - Depolarization

$$\Delta E_{\text{depol}} = (E_{\text{off}} - E_{\text{depol}}), \tag{6.10}$$

where

ΔE_p	polarization for criterion (100-mV minimum)
E_{off}	potential with all current momentarily interrupted (millivolts)
E_{native}	native (free corroding) potential before cathodic protection current is applied (millivolts)
ΔE_{depol}	depolarization for criterion (100-mV minimum)
E_{depol}	depolarized potential with current left off (millivolts)

7.0 REFERENCES

7.1 NACE Standard SP0177-2007, "Mitigation of Alternating Current and Lightning Effects on Metallic Structures and Corrosion Control Systems" (Houston, TX: NACE International, 2007).

7.2 NACE Standard SP0169-2007, "Control of External Corrosion on Underground or Submerged Metallic Piping Systems" (Houston, TX: NACE International, 2007).

7.3 CGA Recommended Practice OCC-1-2005, "For the Control of External Corrosion on Buried or Submerged Metallic Piping Systems" (Canadian Gas Association, Ottawa, Ontario, 2005).

7.4 ISO 15589-1, "Petroleum and Natural Gas Industries: Cathodic Protection of Pipeline Transportation Systems, Part 1—On-Land Pipelines" (Geneva, Switzerland: ISO, 2003).
7.5 NACE Standard TM 0497-2002, "Measurement Techniques Related to Criteria for Cathodic Protection on Underground or Submerged Metallic Piping Systems" (Houston, TX: NACE International, 2002).
7.6 A.W. Peabody, *Control of Pipeline Corrosion*, 2nd ed., ed. R.L. Bianchetti (Houston, TX: NACE International, 2001), pp. 72, 126.

CATHODIC PROTECTION PROCEDURE NO. 8

Direct Current Stray Current Interference 8

1.0 INTRODUCTION

The purpose of this procedure is to investigate the possibility of basic direct current (DC) stray current interference on underground or submerged structures. Complex dynamic interference problems such as those encountered in transit system operations are not included in this procedure as each situation is somewhat unique. A basic understanding of dynamic interference is included.

DC stray current can be defined as a current in an electrical path other than the intended circuit. Corrosion occurs where a conventional DC current leaves a metal to enter an electrolyte such as water or soil; therefore, if a stray DC current leaves a structure to enter the electrolyte, it accelerates corrosion that is in proportion to the amount of current and the time of exposure. A DC current discharge from a metal into an electrolyte, such as illustrated in Figure 1.1, causes corrosion at the following rates for these metals[(1)]:

Iron	9.2 kg/A-year	(20.1 lb/ampere-year)
Copper	20.7 kg/A-year	(45 lb/ampere-year)
Lead	33.9 kg/A-year	(74.6 lb/ampere-year)

[1] Developed from Faraday's law, $W = MIt/nF$, where W is weight loss (grams [g]), M is atomic weight of metal (g), I is corrosion current (amperes), t is time of exposure (seconds [s]), n is number of charges transferred in the reaction, and F is Faraday's constant (~96,500 C/equivalent weight of material, and equivalent weight = M/n).

For example, if a current of 1 ampere (A) is discharged from a steel surface into an electrolyte for 1 year, 9.2 kg (20.1 lb) will be lost. Should this current be confined to a small area, then a rapid penetration of the structure wall can be expected.

There also appears to be a relationship between alternating current (AC) and corrosion; however, it is not as well defined as that for DC.[7.1] AC stray current can also cause a hazardous issue.[7.2,7.3] Cathodic protection (CP) appears to control corrosion at a current density less than 20 A_{AC}/m^2 (2 A_{AC}/ft^2). Between 20 A_{AC}/m^2 and 100 A_{AC}/m^2 (10 A_{AC}/ft^2), corrosion is unpredictable, but greater than 100-A_{AC}/m^2 corrosion can be expected regardless of the amount of cathodic protection. Safety is of paramount concern when high AC voltages are present.[(2)]

Sources of stray current include, but are not limited to, the following list:

- CP[7.4]
- High DC voltage (HVDC) power lines[7.5]
- Transit systems[7.6,7.7]
- DC-operated mining equipment
- Electric railways
- Welding[7.8]
- Electroplating equipment with ground faults
- Battery charging equipment with ground faults
- Natural (telluric) current[7.9]
- High AC voltage (HVAC) power lines[7.3,7.10]

With the exception of certain tests on rectifiers, CP Testers are qualified to conduct tests that detect stray current interference but not to mitigate it.[(3)] When detected, the project manager or supervisor must be notified immediately. CP Technicians or equivalent are qualified to complete the field tests to detect and to mitigate basic DC interference effects. CP Technologists and CP Specialists or equivalent are qualified to conduct tests to detect and to mitigate more complex interference effects and to complete the analysis.

---

[2] *Also see* Cathodic Protection Procedure No. 11: Alternating Current Voltage Testing.

[3] *See* Cathodic Protection Procedure No. 1: Rectifier Adjustment, Inspection, and Basic Troubleshooting, for qualifications to work on rectifiers.

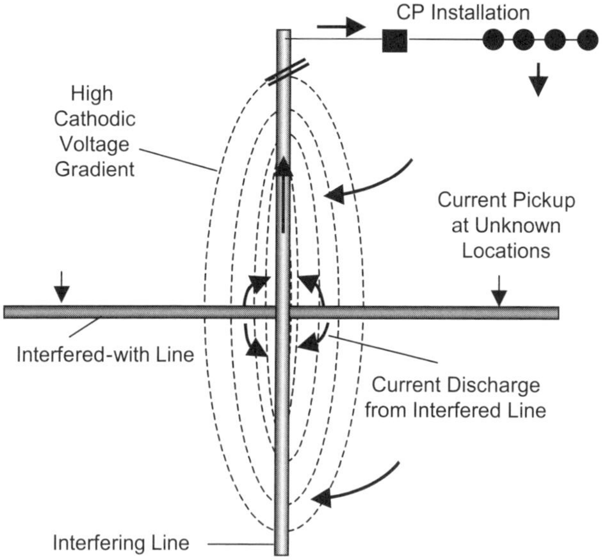

Figure 1.1 Example of cathodic interference is shown.

Complex interference effects include dynamic interference and stray current from several sources.

The CP field personnel must have a good understanding of the criteria as well as the special conditions and precautionary notes that apply to these criteria.[4]

A hazard assessment and a job safety analysis are to be completed for each phase of the project, and the appropriate preventative measures are to be taken.[5]

Static or steady-state stray currents are from sources that are uniform in amplitude and direction and include CP systems and, to some extent, HVDC transmission lines. Dynamic stray current varies in amplitude and often in direction and includes the other sources listed previously.

Cathodic interference occurs when a structure passes through a strong voltage gradient that is negative with respect to remote earth (see Figure 1.1).

[4] *See* Cathodic Protection Procedure No. 2: Structure-to-Electrolyte Measurement.

[5] *See* company or client Safety Manual.

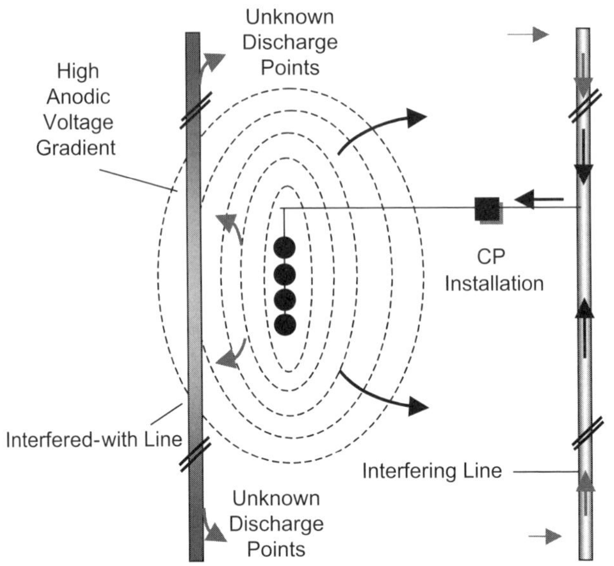

Figure 1.2 Example of anodic interference is shown.

Current is encouraged to discharge from the interfered structure to the soil at this point, thus accelerating corrosion. The corresponding current pickup can take place over an unknown distance away.

Anodic interference occurs when the structure passes through a strong voltage gradient that is positive with respect to remote earth. In this case, current is encouraged on to the structure and then may discharge over an unknown distance to return to its source (see Figure 1.2).

Although not required, anodic and cathodic interference can also take place together.

The following conditions contribute to interference conditions:

- Relative location of the structure to the source of stray DC current
- Magnitude and density of the current
- Effectiveness of the structure's coating
- Proximity to foreign structure
- The environments that the structure passes through: rock versus soil, water, rivers, and so on

- Electrical discontinuities in the structure caused by electrical isolating fittings, sections of line replaced and left in place, and so on
- Telluric current

Telluric current is the naturally occurring current in the earth as a result of geomagnetic fluctuations in the earth, in turn caused by solar storms interfering with the earth's magnetic field extending into outer space. The earth's magnetic field is generally from north to south but does vary throughout the world. Telluric current is normally recognized by slow fluctuations in structure-to-electrolyte potentials that can go from much more negative than normal values to even positive readings.

2.0 TOOLS AND EQUIPMENT

The following equipment will vary, depending on the test selected:

- Multimeter capable of measuring 1 mV_{DC} to 40 V_{DC}, complete with leads with insulated probes
- Data logger(s) capable of measuring 1 mV_{DC} to 40 V_{DC}, complete with leads with insulated probes, preferably Global Positioning System (GPS) time-synchronized
- Copper-copper sulfate reference electrode
- Multimeter including AC/DC volts and an ohmmeter
- Current interrupter(s), preferably GPS time-synchronized
- DC ammeter sized for test current
- Clamp-on ammeters, if appropriate
- Test wires as necessary for the applicable test
- Small hand tools

3.0 SAFETY EQUIPMENT

- Standard safety equipment and clothing, as required by the owner's or your company's safety manual
- Lock-out/tag-out kit
- Electrically insulated clips and probe handles for meter leads

- Only personnel who have received training and are qualified in accordance with local codes and regulations are to work on DC power sources or their supply.

4.0 PRECAUTIONS

The following precautions are in addition to those that must also be followed when working on a particular facility.

4.1 Measure the voltage between the rectifier case and ground before touching the case.

4.2 Open the case expecting to find biting insects, rodents, or snakes inside and take the appropriate precautions.

4.3 Inspect the rectifier for abnormal sounds, temperature, or odors and, if noted, turn it off.

4.4 Switch off and lock out/tag out the AC voltage supply before installing a current interrupter or each time the taps are adjusted.

4.5 Secure any exposed electrical terminals in a locked enclosure when the rectifier is not attended.

4.6 Measure a structure AC voltage to ground on the structure before taking CP measurements. If the AC voltage to ground is equal to or exceeds 15 V_{AC}, practice the safety measures detailed in NACE SP0177[7.3] and advise other personnel working on the structure of the hazard.

4.7 When working near HVAC power lines, take AC structure-to-ground voltage readings at frequent intervals as these voltages can change with the power line load and geometry.

4.8 Do not work on the structure when lightning is in the area.

4.9 When working near a fence, confirm that it is not an electric fence for livestock (look for insulators) and that a parallel HVAC power line is not inducing an AC voltage on it.

4.10 Take precautions for local hazards such as traffic or dangerous animals and insects.

4.11 Only personnel qualified in DC stray current testing and analysis are to be responsible for these tests.

5.0 PROCEDURE

5.1 Initial Detection of Interference (Stray) Current

5.1.1 Suspect the presence of DC stray current under the following conditions, recognizing that the cause could be due to other reasons.

5.1.1.1 Variation of structure-to-electrolyte potentials at one location over time

5.1.1.2 A significant electronegative change in potentials along a structure, especially in close proximity to structures that have CP or have DC-operated equipment

5.1.1.3 An electropositive change in potentials along a structure not near the structure's CP system

5.1.1.4 Abrupt changes in the structure-to-electrolyte potential over a short distance, especially near other facilities with operating DC power sources

5.1.1.5 Structure current changes in magnitude and direction

5.1.1.6 Localized pitting, especially in areas of other DC current sources

5.1.1.7 Localized structure coating damage, especially near DC current sources

5.1.2 Suspect AC stray current with a structure-to-earth voltage greater than 5 V_{AC}.

5.2 Earth Voltage Gradients

5.2.1 Be aware that a change in structure to a reference electrode potential measured at the surface may also occur due to a soil voltage gradient change from current in the soil only and may not be a current pickup/discharge on the structure.

5.2.2 Confirm by measuring side structure-to-electrolyte potential[6] to a minimum 3 m transverse to the pipeline, as shown in Figure 5.1, with the interfering source being interrupted, if possible. Keep the reference electrode spacing on each side equal.

---

[6] *See* Cathodic Protection Procedure No. 2: Structure-to-Electrolyte Potential Measurement.

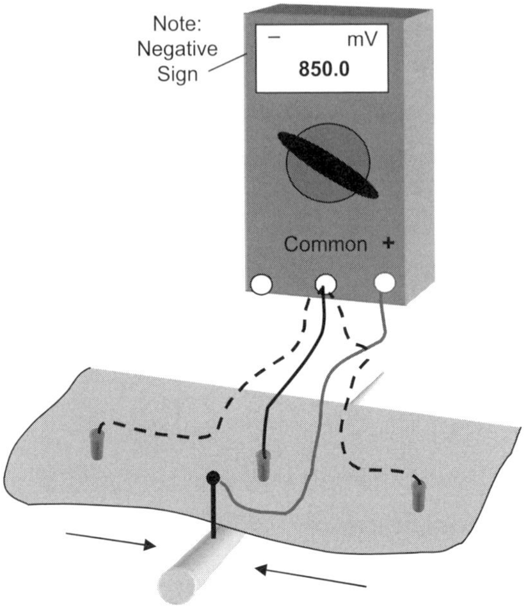

Figure 5.1 Structure-to-electrolyte potentials measured transverse to a pipeline are shown.

5.2.3 Alternately, measure potentials between two reference electrodes transverse to the pipeline, as shown in Figure 5.2.

5.2.4 Be careful to record the correct polarity of each reading using the two-reference electrode method as entirely different conclusions can result with the same current measurements, as shown in the three cases in Figure 5.2. A sample field note format is given in Table 5.1.

5.2.5 Use different reference electrode spacing or leapfrog the reference electrodes to note the strength of the earth voltage gradients.

5.2.6 A grid pattern can also be measured and the gradient contours plotted.

5.2.7 When using specialized equipment to measure earth current, follow the manufacturer's instructions on operation. Select measurement locations with the greatest likelihood of stray current (see Figure 5.3).

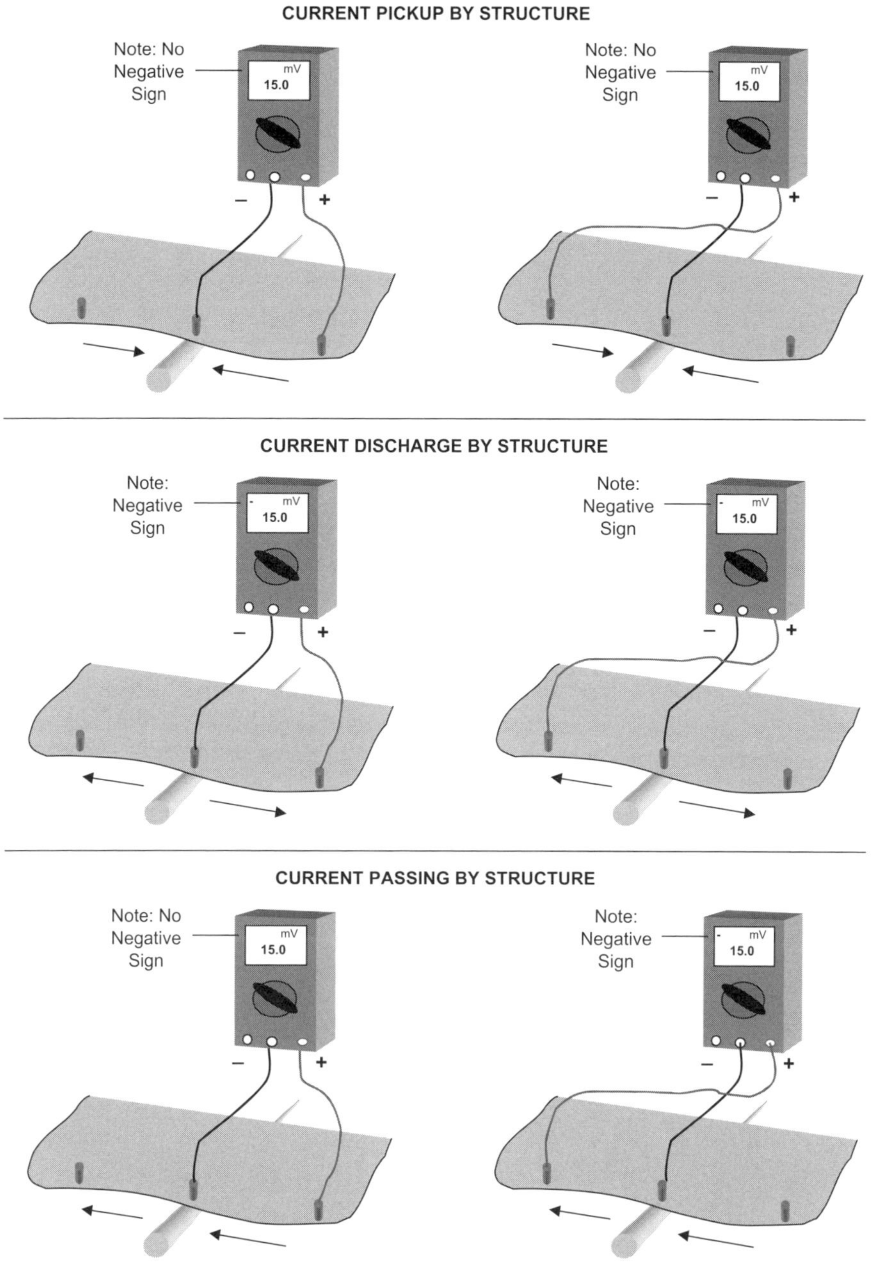

Figure 5.2 Potentials measured between two reference electrodes transverse to the pipeline are shown. Note that polarity is critical in the analysis.

Table 5.1 Sample Field Notes for Two Reference Electrode Voltage Gradient Measurement

	Structure-to-Electrolyte Potentials (units)								
	North/West			Center Reference Positive or	South/East			Center Reference Positive or	Reference
Location	On	Off	Net IR_1	Negative?	On	Off	Net IR_2	Negative?	Spacing

5.3 Steady-State Anodic Interference

5.3.1 Suspect anodic interference if there is an electronegative increase in potentials near a foreign or interfering DC current source, as illustrated in the potential profile example in Figure 5.4.

5.3.2 If the interfering source is relatively steady, arrange to have the interfering source either interrupted on a timed cycle or temporarily turned off.

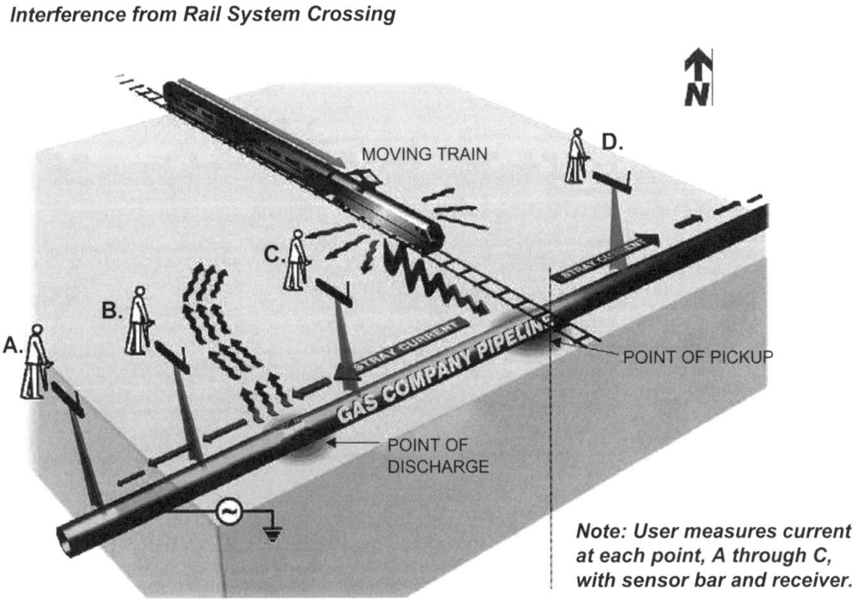

Figure 5.3 Earth current measurement is shown.[7.11]

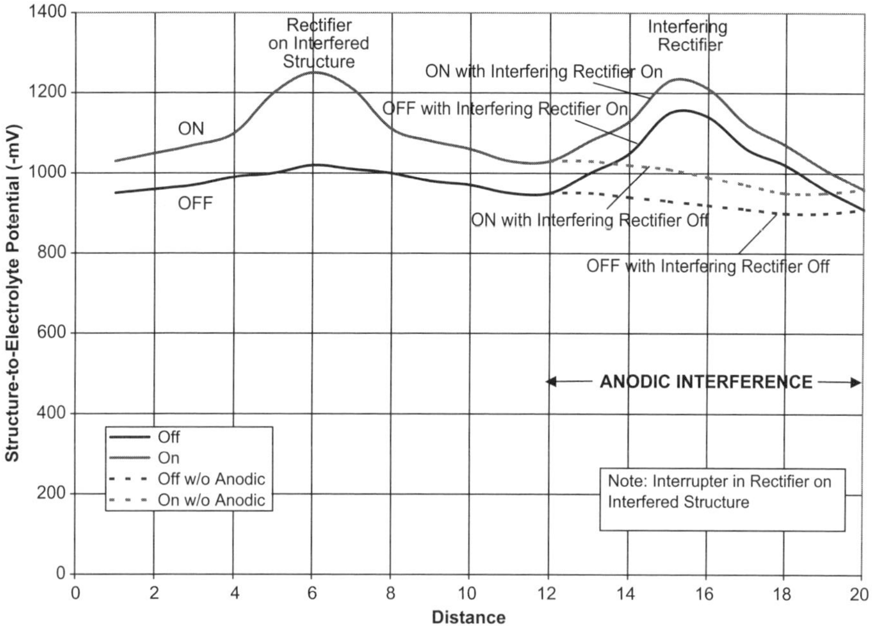

Figure 5.4 Example of anodic interference is shown.

5.3.3 Conduct a close interval survey (CIS)[(7)] to define the anodic gradient or current pickup area. This area is defined where the structure-to-electrolyte potential is driven more electronegative by the interfering (foreign) DC power source.

5.3.4 A corresponding current discharge area exists, and if it should be entering the electrolyte, then it must be located; otherwise, a corresponding amount of current must be drained from the structure by means of a metallic path.

5.3.5 Take test station readings, especially at any points where the interfering structure comes close, to note any current discharge points.

5.3.6 Where current discharge is noted, conduct a CIS to define the point(s) of greatest exposure.

5.3.7 At these locations, complete the tests outlined in Section 5.4.

[7] *See* Cathodic Protection Procedure No. 7: Close Interval Potential Survey.

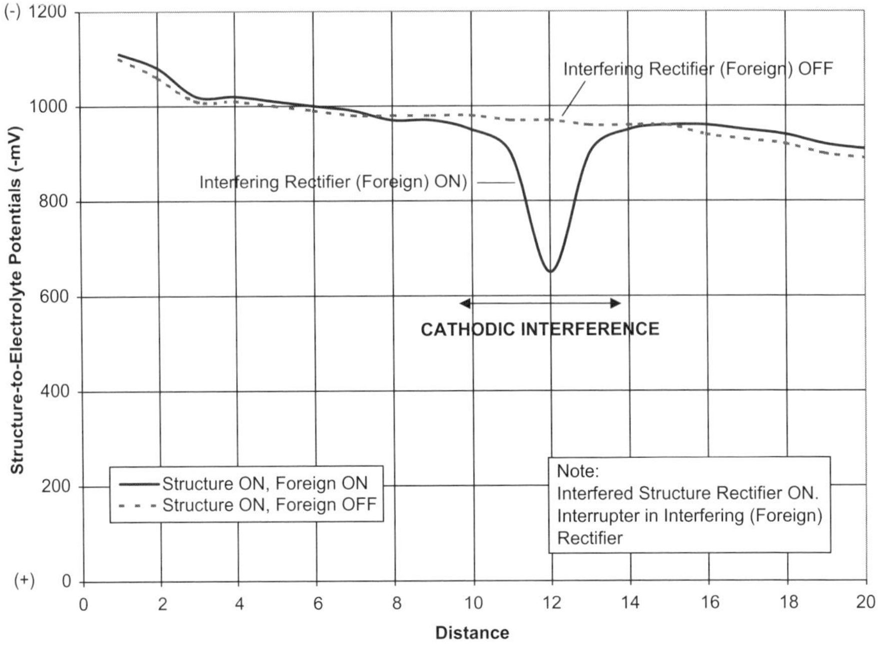

Figure 5.5 Example of cathodic interference is shown.

5.4 Steady State Cathodic Interference

5.4.1 Suspect cathodic interference if there is an electropositive change in potentials, especially near a foreign or possible interfering structure, as illustrated in the ON potential profile example in Figure 5.5.

5.4.2 If the interfering source is relatively steady, arrange to install an interrupter on a known timed cycle in the interfering DC power source.

5.4.3 Conduct a CIS[(8)] to define the cathodic gradient or current discharge area. This area is defined as where the potential is driven more electropositive by the interfering (foreign) DC power source.

---

8 *See* Cathodic Protection Procedure No. 7: Close Interval Potential Survey.

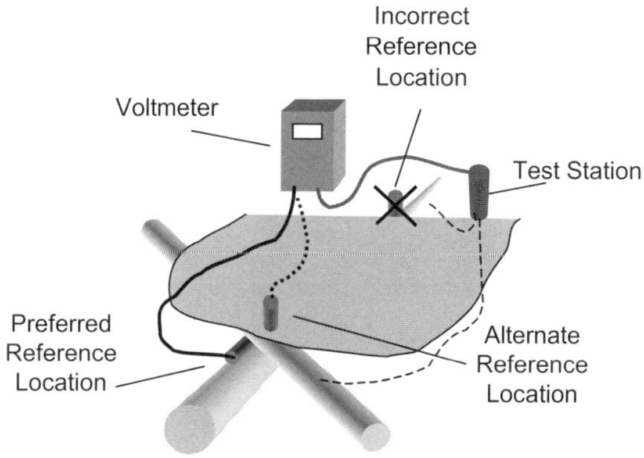

Figure 5.6 Preferred reference electrode positions at a pipeline crossing are shown.

5.4.4 Take additional CIS tests, especially at any points where the interfering structure comes close to the interfered structure, to note any other current discharge points.

5.4.5 Where current discharge is noted by a more electropositive shift in potentials, expand the CIS to define the point(s) of greatest exposure.

5.4.6 When measuring structure-to-electrolyte potentials at a point of maximum exposure, such as at a pipeline crossing, the position of the reference electrode is critical. Locate the reference electrode as close as possible to the point of maximum exposure on the interfered structure, rather than indifferently near the test station or point of contact to the structure, as shown in Figure 5.6. Otherwise, the measurement will include earth gradients (*IR* drop), with the pipe acting as the second reference electrode, and the readings will include a large error.

5.4.7 Consider installing a stationary reference electrode at the point of maximum exposure for more accurate testing.

5.4.8 If the test station is situated at a correct position to test for stray current measurements, the reference electrode can be placed inside the test station, as shown in Figure 5.7. The potential measured will be the same as if the electrode were placed at the bottom of the test post. The soil in the test post forms an

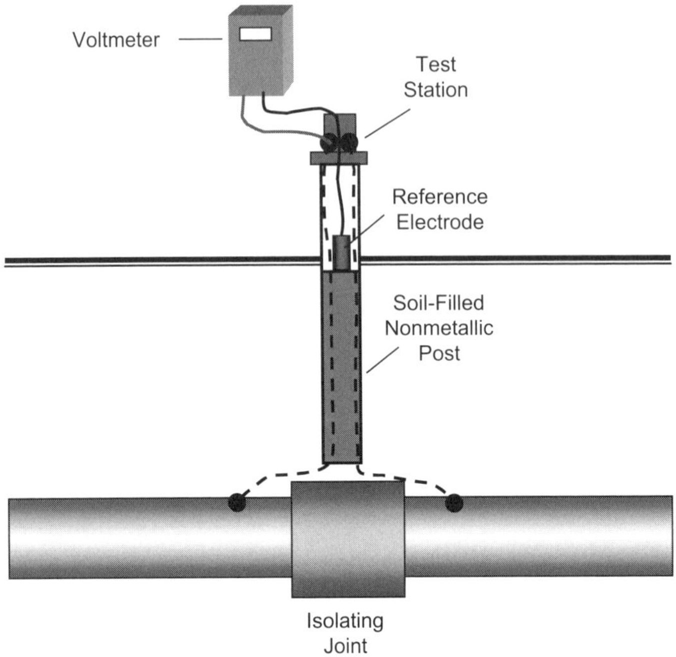

Figure 5.7 Placement of a reference electrode inside a nonmetallic post to measure stray current at an isolating joint.

electrolytic bridge between the reference electrode and the bottom of the test post, or one might look at it as an extension of the reference electrode porous plug.

5.4.9 If the test station post is metallic or in the wrong location, consider installing a nonmetallic pipe to the correct point and filling it with native soil. The reference electrode is then placed inside the nonmetallic pipe.

5.4.10 Where earth voltage gradients are interfering with the true structure-to-electrolyte potentials, consider installing a coupon-type test station, as shown in Figure 5.8.

5.4.11 Use the coupon for two purposes: first, measure a current pickup or discharge with a zero-resistance ammeter, and second, measure true structure-to-electrolyte potentials with the reference electrode inside the nonmetallic test post (see Figures 5.6 and 5.7).

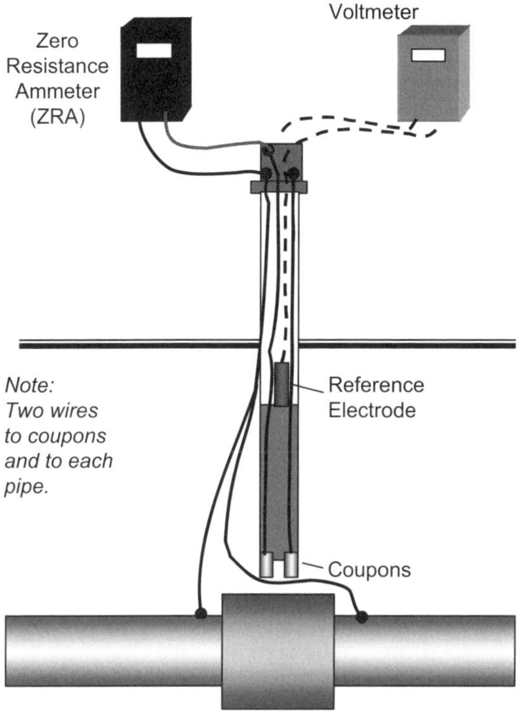

Figure 5.8 Use of coupon test station for DC stray current measurement is shown. Adjust bottom of post to critical point for stray current.

5.5 Current Mapping

Current mapping normally applies to linear structures, such as a pipeline, where the current can be measured at regular intervals.

5.5.1 Measure the current in the pipeline or structure at intervals, where possible, by using a current span,[(9)] a clamp-on ammeter, or a Pipeline Current Mapper™at regular intervals along the pipeline, as shown in Figure 5.9.

5.5.2 Record both the pipeline current and direction. An example is given in Figure 5.9, where a positive current is in the same direction as product flow and a negative value is given to current in the opposite direction. (Continuously increasing current

[9] *See* Cathodic Protection Procedure No. 3: Direct Current Measurements.

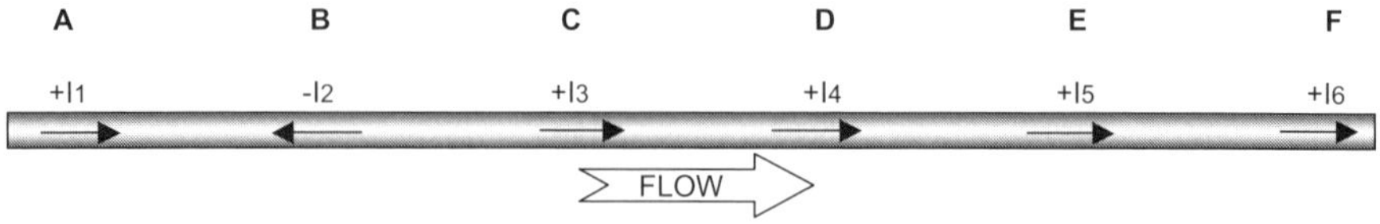

Figure 5.9 Pipeline current polarity convention is shown.

indicates a current pickup, whereas decreasing current or current in opposite directions indicates a current discharge.)

5.6 Dynamic Interference

5.6.1 Dynamic interference is a stray current that varies in amplitude, and often in direction.

5.6.2 Determine the source of stray current. Rapid changes in potential with a repeatable pattern and quiet periods suggest a man-made stray current. Slow variations without a noticeable pattern may be a natural or telluric current.

5.6.2.1 Observe the facilities that are in the area that may be a source of dynamic interference.

5.6.2.2 Data log the structure-to-electrolyte potentials over a period of 22–24 hours to see if a wave print of the potentials will help identify the source. Should quiet periods appear at specific times of day, attempt to relate them to normal work shifts at a plant or mine.

5.6.3 If the stray current is from a man-made stationary source, install a data logger to monitor the structure-to-electrolyte potentials and, if practical, structure current at suspected locations of exposure. The data loggers must be time-synchronized.

5.6.4 If the stray current is from a man-made moving source, install data loggers at several points along the structure to measure the structure-to-electrolyte potential and the structure current simultaneously. The data loggers must be time-synchronized.

5.6.5 If the stray current is telluric, install two stationary data loggers at the one-fourth and three-fourths points of the section under test. A single data logger may monitor a short section under test. Take each measurement for approximately 5 min with a portable data logger that is time-synchronized to the stationary data loggers.

5.7 Internal Interference Current

5.7.1 Internal interference usually occurs when a stray current bypasses an isolating feature through a low-resistivity electrolyte inside the pipe.

5.7.2 Measure structure-to-electrolyte potentials with a suspected source being interrupted. A current pickup is indicated by an electronegative change in potentials.

5.7.3 An electropositive shift in potentials, however, will not be recorded at the internal current discharge point with the reference electrode in the electrolyte (soil or water) outside the pipe.

5.8 Direct Current Stray Current Testing After Mitigation

5.8.1 Use a similar test procedure as that used to detect DC stray current to ensure that stray current effects are mitigated, with the recognition that conditions will have changed, depending on the mitigation technique, as described subsequently.

5.8.2 Bonds

5.8.2.1 These will normally be defined as a critical bond and in regulated pipelines must be inspected bimonthly.

5.8.2.2 The structure-to-electrolyte potentials being measured with two structures bonded are a mixed potential between the two structures and will vary depending on the resistance of the bond.

5.8.2.3 The change in potential from the ON to the OFF structure-to-electrolyte potential on the interfered structure with the interfering rectifier(s) being interrupted does not necessarily reflect interference. Instead, the interfered structure potential may be going more electropositive as it is now trying to protect the interfering (foreign) structure when the interfering CP system is interrupted.

5.8.2.4 Normally, the ON potential of the interfered structure, when bonded, is brought back to the ON potential, with the interfering current off and no bond.

5.8.2.5 Install synchronized interrupters in the influencing interfering rectifier(s), interfered rectifier(s), and the bond to obtain a true polarized (OFF) potential on either structure. Different cycles can be used to distinguish between the two CP systems, but they must overlap during the ON and OFF cycles.

5.8.2.6 Measure the ON and OFF structure-to-electrolyte potentials on both structures to ensure that the interfered structure potentials have been brought back to their normal values.

5.8.3 Sacrificial Anodes

5.8.3.1 Although not a critical bond, these should also be inspected frequently as the consequences of their failure may be severe.

5.8.3.2 The structure-to-electrolyte potentials being measured with a sacrificial anode(s) attached will be a mixed potential between the sacrificial anode(s) and the structure. With the interfering rectifier on, the magnesium anode will discharge current, but that current is intended to return to the interfering structure. An electropositive change in potentials may therefore be seen due to this discharge when the interfering rectifier goes on. This change is not a reflection of interference to the structure but of the intended current discharge from the sacrificial anode, and suggests that the reference electrode is not in the correct location.

5.8.3.3 Locate the reference electrode at the point of structure current discharge without the sacrificial anodes connected and measure the ON/OFF potentials with the interfering rectifier interrupted.

5.8.3.4 Measure the ON/OFF potentials with the sacrificial anodes connected. If the change is in a more electronegative direction when the interfering rectifier comes on, then document the results. If the change is in a electropositive direction, move the reference electrode without moving it away from the cathodic gradient at the structure to determine if it is in the anodic gradient of the sacrificial anode. If the structure-to-electrolyte potential is still going more electropositive during the ON cycle of the interfering rectifier, then additional anodes will be required.

5.8.3.5 Measure the current and current direction of the sacrificial anodes.

5.8.3.6 Install synchronized interrupters in the influencing interfering rectifier(s), interfered rectifier(s), *and* the sacrificial anode(s) to obtain a true polarized (OFF) structure-to-electrolyte potential on the interfering structure. Different cycles can be used to distinguish between the two CP systems, but they must overlap during the ON and OFF cycles.

5.8.3.7 Measure the ON and OFF structure-to-electrolyte potentials on both structures and ensure that the interfered structure potentials have been brought back to normal. (Note that if they were not protected before, they do not need to be protected now, but only brought back to their original potential.)

5.8.4 Additional Cathodic Protection

5.8.4.1 In addition to installing sacrificial anodes, an impressed current cathodic protection (ICCP) system can be installed to balance cathodic gradients between the structures.

5.8.4.2 Conduct the tests above before the installation of additional ICCP and determine if the interfering or interfered structures are changed.

5.8.4.3 Adjust the new ICCP system such that there is no interference on either structure.

5.8.5 Recoating Current Pickup Areas

5.8.5.1 Note that coating the current discharge areas can concentrate the interference current at smaller coating holidays, thus causing a faster penetration of the structure wall. Coating the current pickup areas is intended to increase the resistance of the path and thus reduce the current pickup in this area. A significant amount of the structure picking up current must be coated.

5.8.5.2 Complete the same tests as before recoating to determine the effectiveness of this solution.

5.8.6 Installing Additional Electrical Isolation

5.8.6.1 Note that the intent of installing additional isolation is to limit long line stray current. The disadvantages are that each section must have an independent CP system, and there is a possibility of creating interference at each isolating feature.

5.8.6.2 Conduct the same tests as before installing isolation, and in addition, complete these tests at each new isolating feature.

5.8.7 Forced Current Bonds

Forced current bonds control the current in a mitigation bond under dynamic stray current conditions.[7.12] This is the result of a complex study that is beyond the scope of this document.

6.0 ANALYSIS

6.1 Earth Gradients

6.1.1 The direction of conventional current is from the positive (+) reference electrode to the negative (−) electrode. If a current is toward the structure on both sides of the structure, then a current pickup is indicated. A current direction away from the structure on both sides indicates a current discharge. A current going toward the structure on one side and away on the other side suggests that the current in the soil is passing by.

6.1.2 Table 6.1 illustrates an example of current pickup, current discharge, and current passing by a pipeline.

6.1.3 The data in this table are from a single pipeline. Analysis of the data from parallel pipelines becomes more complex as the information from each pipeline must be overlapped.

6.2 Steady State Anodic Interference

6.2.1 Anodic interference is often first suspected by a sudden electronegative change in structure-to-electrolyte potentials in an area where there is no CP anode bed for the structure (see Figure 5.4).

Table 6.1 Sample Side Drain Potential Field Notes

	Structure-to-Electrolyte Potential (mV)						
	North of Pipeline (3m)		Over Pipeline		South of Pipeline (3m)		
Location	OFF	ON	OFF	ON	OFF	ON	Remarks
MP 10	−960	−1500	−940	−1300	−980	−1550	Current Pickup
MP 20	−800	−650	−820	−750	−800	−680	Current Discharge
MP 30	−920	−1600	−940	−1700	−960	−1800	Current Passing by Pipeline

6.2.2 When the DC power source not connected to the structure is interrupted, an electronegative shift in the structure-to-electrolyte potentials on the interfered-with structure indicates anodic interference on that structure. Although this indicates a current pickup, there must be a corresponding current discharge at another location(s) on the structure for the current to return to its source. The latter point(s) are of concern as corrosion is accelerated at these locations.

6.2.3 Investigate the possibility of moving the offending anode bed or electrical ground so that the structure is no longer in the anodic gradient.

6.2.4 Otherwise, locate the current discharge locations and, if through the soil, complete the cathodic interference analysis discussed in Section 6.3.

6.2.5 As a temporary measure, a current can be drained back to the source through a bond. A unidirectional device may be installed in this bond to prevent a reverse of current, if so desired.

6.2.6 An estimation of the amount of current necessary can be made by first calculating the *IR* drop due to the current pickup. Then, drain a current to the pipeline from a temporary ground located near the offending anode bed that will add the equivalent amount of drop to the first. Drain this current as a temporary measure, until further investigations can be made into the current discharge locations.

6.3 Steady State Cathodic Interference

6.3.1 Cathodic interference is often first detected by a more electropositive change in potentials in a structure, especially where close to another electrically isolated structure. This is confirmed when the DC power source for the electrically isolated structure is interrupted and the structure-to-electrolyte potential on the interfered-with structure becomes more electronegative when it is turned off and more electropositive when it is on (see Figure 5.5).

6.3.2 A current discharge occurs when the interfered-with structure becomes more electropositive when the interfering current source comes on. Either the source of this current must be eliminated, or it must be safely returned to the original source.

6.3.3 Elimination of the interfering current might be accomplished by the following:

- 6.3.3.1 Reducing the output of the offending DC power source
- 6.3.3.2 Correcting a fault to ground in a DC power source that only uses the earth during a fault
- 6.3.3.3 Recoating the current pickup area to reduce the cathodic gradient and increase the resistance of the circuit
- 6.3.3.4 Potentially controlled rectifiers

6.3.4 Returning the interfering current safely to its source may be accomplished by the following:

- 6.3.4.1 Mitigation bond
- 6.3.4.2 Sacrificial anodes
- 6.3.4.3 Additional impressed current CP
- 6.3.4.4 Forced current bonds

6.3.5 Combinations of the preceding solutions can also be used.

6.4 Current Mapping

6.4.1 Identify current pickup and discharge locations by a summation of the current in the pipeline.

- 6.4.1.1 The sum of the current at any current junction must equal zero or the current into a point must equal the current leaving that point.
- 6.4.1.2 An example is given in Figure 6.1. From location A to location B, the current changes from 1.0 A to 2.0 A, but

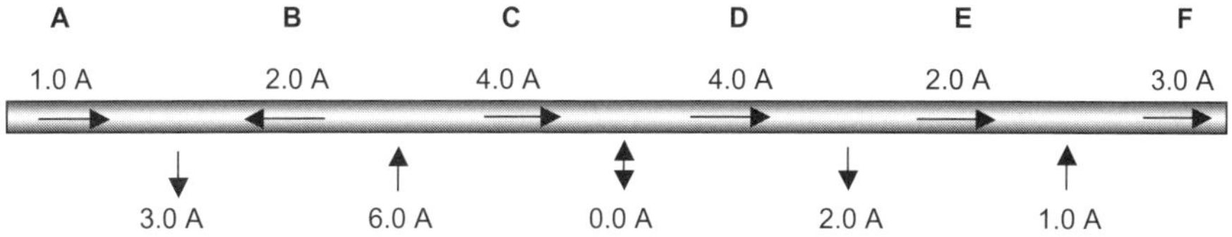

Figure 6.1 Current mapping of a pipeline section is shown.

the current at B is in the opposite direction (recorded as −2.0 A); therefore, a current discharge of 3.0 A took place in between the pipeline measurements (A to B). The current at location C changed to +4.0 A; therefore, a current pickup of 6.0 A occurred between locations B and C. There was no change in pipeline current at locations C and D; therefore, there was no current pickup or discharge between these locations. The current at location E is in the same direction as at location D but is less; therefore, a current discharge must have taken place in between these locations (D to E). The current at location F is in the same direction but greater than E; therefore, current pickup occurred in this span (E to F).

6.4.2 Plot these values on a graph, as shown in Figure 6.2, where the current going in the opposite direction is given a negative value. The current pickup and discharge sections of pipe can readily be noted by the pipeline current profile, where a positive slope indicates a current pickup and a negative slope indicates a current discharge.

6.4.3 Note the sections of current discharge for remedial action.

6.5 Dynamic Interference

The assessment of dynamic interference is complex, as are the solutions, and is beyond the scope of this document. Some brief comments are offered subsequently.

6.5.1 Telluric (Natural) Stray Current

6.5.1.1 Although telluric currents can cause an interference problem, the main difficulty is the fact that the actual structure-to-electrolyte potential is not obvious; thus, any changes along the route of a structure can be masked.

Figure 6.2 Pipeline current plot with current direction left to right is shown. Current pickup is indicated by a positive slope and current discharge by a negative slope.

6.5.1.2 There are several methods that can be used to determine the actual structure-to-electrolyte potential under telluric conditions, one of which is described in "Cathodic Protection Procedure No. 2: Structure-to-Electrolyte Potential Measurement," Section 6.5.

6.5.2 Man-Made Dynamic Stray Current from a Stationary Source

6.5.2.1 A stationary source of dynamic current is assessed in increments of time in a similar manner as for steady state interference. The periods of current discharge and the length of that discharge are of major concern.

6.5.2.2 The results after mitigation should minimize the structure-to-electrolyte variations and keep them at a structure-to-electrolyte potential equivalent or more electronegative than without the interference.

6.5.3 Man-Made Dynamic Interference from a Moving Source

6.5.3.1 The conditions at any one point on the structure are continually changing, with the magnitude of the stray current varying in addition to the location of the source.

6.5.3.2 If a bond is used to return the current to its source, a major concern is the location of this bond. The preferable location is at the natural point of discharge. Locating it at another point may simply be providing a parallel path to the current discharge point and, therefore, will not solve the overall problem.

6.5.3.3 One method that is used to locate the general current pickup and discharge locations is often called the "beta curve" technique. The method compares the current from one location to the next location in a plot and also compares the line current to the structure-to-electrolyte potential at each location in another plot. Knowing the current direction, the slope of the plot indicates a general current pickup or discharge.

6.5.3.4 Another method is to compare the data logs of the line current, structure-to-electrolyte potentials, and, in the case of a transit system, rail-to-structure potentials at each location.

6.5.3.5 Similar procedures are used after the mitigation is in place.

6.6 Earth Current

6.6.1 Identify current pickup and discharge areas in a similar manner as current mapping (see Figure 6.3).

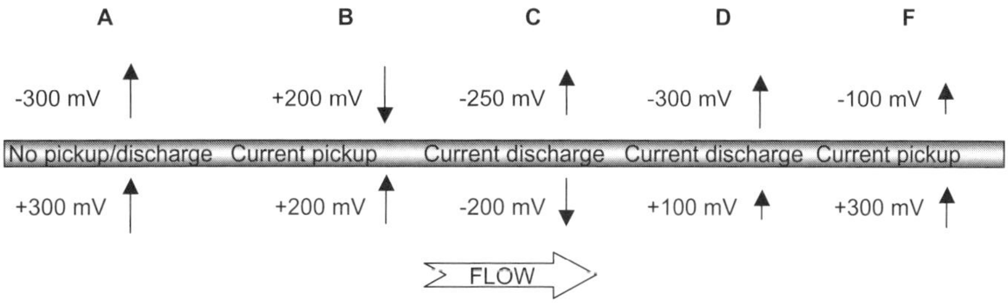

Figure 6.3 Stray earth current mapping is shown.

6.6.2 Although a quantitative measurement of earth current is not intended, a relative current can be estimated by a direct comparison of potentials, provided the soil resistivity and the reference electrode spacing remain the same.

6.6.3 The change in current will be inversely proportional to the change in resistivity.

7.0 REFERENCES

7.1 R.A. Gummow, R.G. Wakelin, S.M. Segall, "AC Corrosion—A New Challenge to Pipeline Integrity," CORROSION/98, paper no. 566 (Houston, TX: NACE International, 1998).

7.2 D.L. Caudill, K.C. Garrity, "Alternating Current Interference—Related Explosions of Underground Industrial Gas Piping," *MP* 37, 8 (1998): pp. 17–22.

7.3 NACE Standard SP0177-2007, "Mitigation of Alternating Current and Lightning Effects on Metallic Structures and Corrosion Control Systems" (Houston, TX: NACE International, 2007).

7.4 A.W. Peabody, *Control of Pipeline Corrosion*, 2nd ed., ed. R.L. Bianchetti (Houston, TX: NACE, 2001), p. 211.

7.5 A.L. Verhiel, "HVDC Interference on a Major Canadian Pipeline Counteracted," *MP* 11, 3 (1972): p. 37.

7.6 P. Frank, "A Review of Stray Current Effects on a Gas Transmission Main in the Boston, Massachusetts Area," CORROSION/94, paper no. 590 (Houston, TX: NACE, 1994).

7.7 W. Sidoriak, "D.C. Transit Stray Current Leakage Paths—Prevention and/or Correction," CORROSION/94, paper no. 585 (Houston, TX: NACE, 1994).

7.8 J.N. Britton, "Stray Current Corrosion During Marine Welding Operations," *MP* 30, 2 (1991): p. 30.

7.9 D.H. Warnke, W.B. Holtsbaum, "Impact of Thin Film Coatings on Cathodic Protection," *Proc. Int. Pipeline Conf.*, paper no. IPC02-27325 (Calgary, Alberta, Canada: ASME, Sept. 29–Oct. 3, 2002).

7.10 NACE International SP0169-2007, "Control of Corrosion on Underground or Submerged Piping Systems" (Houston, TX: NACE International, 2007).

7.11 A. Kacicnik, D.H. Warnke, G. Parker, "Stray Current Mapping Enhances Direct Assessment (DA) of an Urban Pipeline," NACE International Northern Area Western Conference (Houston, TX: NACE International, Feb. 2004).

7.12 J.I. Munroe, "Optimization of Reverse Current Switches," CORROSION/80, paper no. 142 (Houston, TX: NACE International, 1980).

CATHODIC PROTECTION PROCEDURE NO. 9

Electrical Isolation 9

1.0 PURPOSE

The purpose of this procedure is to describe tests to determine the effectiveness of aboveground or underground electrical isolating (insulating) features and to determine electrical continuity between structures.

2.0 TOOLS AND EQUIPMENT

The following equipment will vary, depending on the test selected:

- Multimeter capable of measuring 1 mV_{DC} to 40 V_{DC}, complete with leads with insulated probes
- Copper-copper sulfate reference electrode
- Isolation checker
- High-voltage resistance meter
- Ohmmeter
- Current interrupter
- Direct current (DC) ammeter sized for test current
- Battery and control resistor or portable controlled DC power supply (6–12 V_{DC})
- Sonic transmitter and receiver
- Test wires as necessary for the applicable test
- Small hand tools

3.0 SAFETY EQUIPMENT

- Standard safety equipment and clothing, as required by the company's safety manual and regulations
- Electrically insulated clips and probe handles for meter leads

4.0 PRECAUTIONS

The following precautions are in addition to those that are required for a particular facility.

4.1 Cathodic protection (CP) systems are designed to protect a specific structure, and as a result, they are electrically isolated from other structures that are not intended to receive CP current from this source.[7.1] CP to the intended structure becomes ineffective when an added current drain is connected through a low-resistance metallic contact in an isolation feature often called a "short" or an "electronic short" (Figure 4.1[A]).

4.2 A change in the structure-to-electrolyte potential on the foreign structure does not necessarily mean that isolation is shorted as it could be a result of stray current.[(1)]

4.3 Note that the reference electrode is to remain at the same location for structure-to-electrolyte potentials intended to determine electrical continuity.[7.2]

4.4 Note that a facility may be shorted through a metallic bypass around the isolating feature, rather than a faulty isolation itself.[7.3] Some common bypasses are shown in Figure 4.1(B).

4.5 An alternating current (AC) voltage can also exist across an isolating feature. Measure an AC structure-to-ground voltage before taking other measurements on the structure. If the AC voltage to ground is equal to or exceeds a hazardous voltage of 15 V_{AC}, proceed in a manner described in NACE SP0177[7.4] and advise other personnel working on the structure of the hazard.

4.6 The voltmeter for structure-to-electrolyte potentials is to be capable of measuring negative values, allowing the negative lead of the

---

[1] *See* Cathodic Protection Procedure No. 8: Direct Current Stray Current Interference.

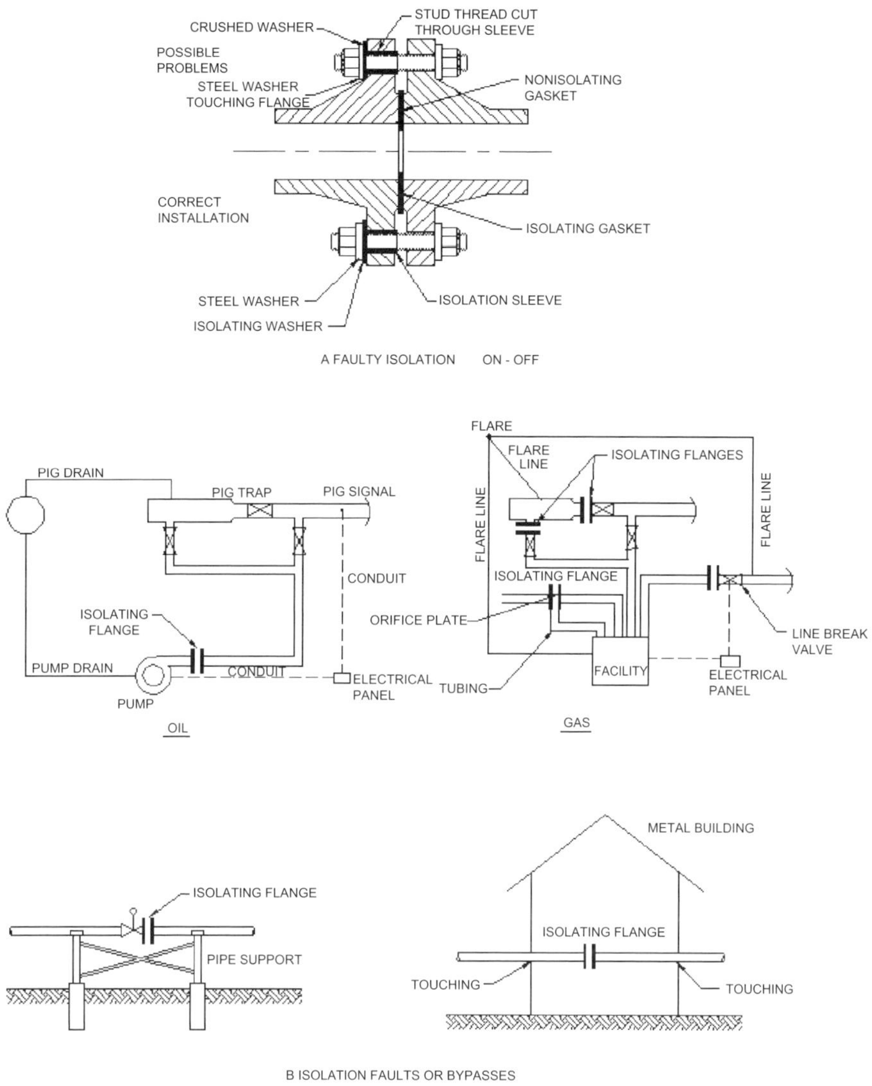

Figure 4.1 Typical isolation bypasses are shown.

voltmeter to be connected to the copper-copper sulfate reference electrode and the positive voltmeter lead to the appropriate test lead and/or structure. When the meter test leads are connected in this manner, the voltmeter will display a negative meter polarity, a potential value, and the units in millivolts or volts. If the meter is not capable of measuring a negative reading, the leads may be reversed, but the value must still be recorded as a negative value.

4.7 Care must be used while pressing the isolation checker pins onto the structure so that they are not broken or pushed into the tester's hands by pushing too hard. The probes on an isolation checker must be in good contact with the pins; otherwise, the instrument will see the isolation as effective, regardless of its condition.

5.0 PROCEDURE

One or more of the following tests can be completed to confirm that the pipeline is electrically isolated from a foreign structure. *Foreign* in this case is a structure that is not intended to receive CP current from this source. It may or may not be the property of the same owner. A flow chart to assist in decision making on the testing process is shown in Appendix A.

5.1 As-Found Pipe-to-Electrolyte Potential Method

5.1.1 For an aboveground isolation, measure a structure-to-electrolyte potential[2] on the pipeline or structure with CP and another on the foreign pipeline or foreign structure *with the reference electrode remaining at the same location* (see Figure 5.1). Alternately, measure a potential difference between the pipe with CP and the foreign pipe directly, as shown in Figure 5.2. This value should equal the difference in the structure-to-electrolyte potential taken on each side with the same reference electrode position.

5.1.2 For an underground isolation, measure a structure-to-electrolyte potential[3] on the pipeline or the structure with CP and another on the foreign pipeline or foreign structure *with the reference electrode remaining at the same location* (see Figure 5.3). Alternately, measure a potential difference between the pipe with CP and the foreign pipe directly, similar to Figure 5.2, but using the test station wires.

5.1.3 Sample field notes are given in Table 5.1.

---

[2] *See* Cathodic Protection Procedure No. 2: Structure-to-Electrolyte Potential Measurement.

[3] *See* Cathodic Protection Procedure No. 2: Structure-to-Electrolyte Potential Measurement.

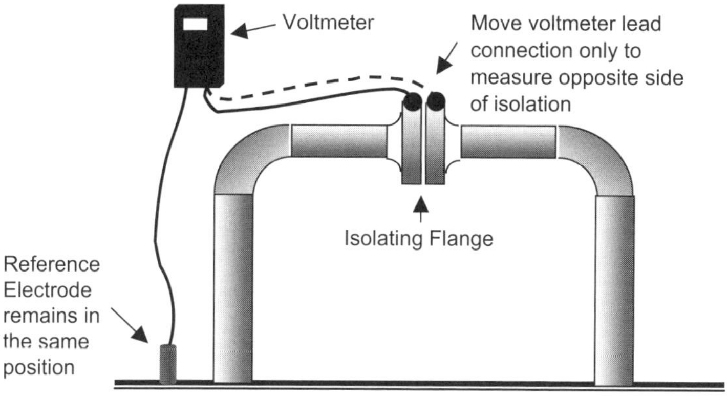

Figure 5.1 Structure-to-electrolyte potential method to test an aboveground isolator is shown.

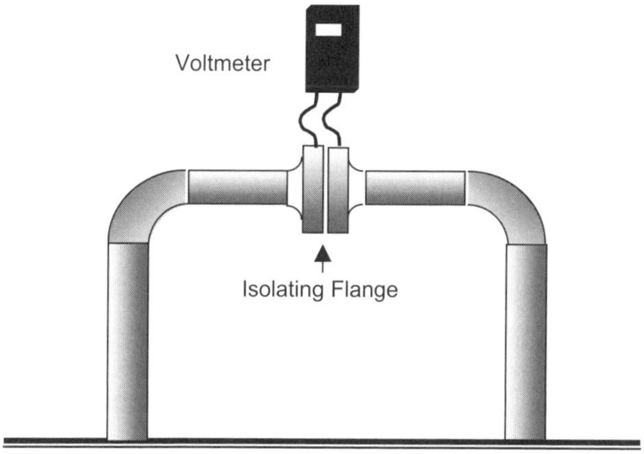

Figure 5.2 Direct voltage measurement across the isolator is shown.

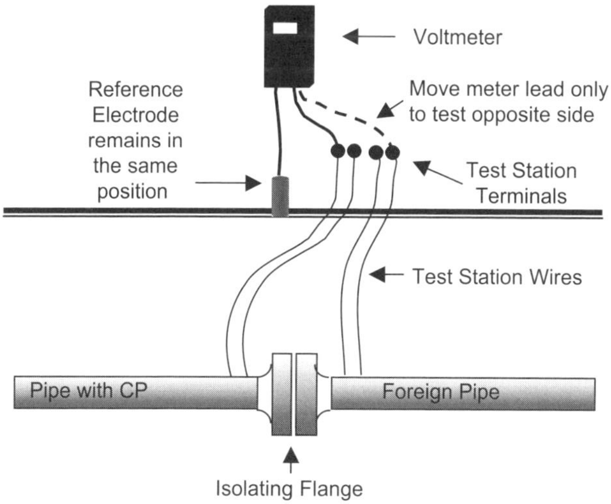

Figure 5.3 Direct potential method to test an underground isolation is shown.

Table 5.1 Typical Field Notes for Direct Potential Method

	Structure-to-Electrolyte Potential (−mV)			
Location	Structure	Foreign	Structure-to-Foreign Potential (−mV)	Remarks

5.2 Interrupted Current Source (ON/OFF Potential) Method

The objective of this method is to change the structure-to-electrolyte potential on one structure that will not be the same on the foreign structure if the two are isolated. Note that this test will indicate whether the structure with CP is either isolated from or shorted to the foreign structure overall, but it will not determine which isolation is shorted if there are more than one.

5.2.1 Interrupt one of the existing CP DC power sources, as shown in Figure 5.4.

5.2.2 Alternately, apply a supplemental interrupted CP current from a temporary DC power source and ground a sufficient distance away so that the isolation is not within the voltage gradient of the temporary anode, as shown in Figure 5.5. If the isolator is aboveground, a similar test setup as in Figures 5.4 and 5.5 still applies, except that the method of measuring structure-to-electrolyte potentials will be that in Figures 5.1 and 5.2.

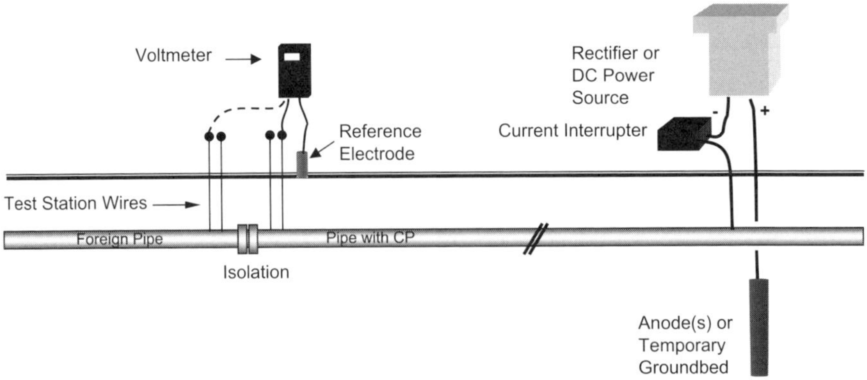

Figure 5.4 Interrupted potential test to determine isolation effectiveness with CP interrupted is shown.

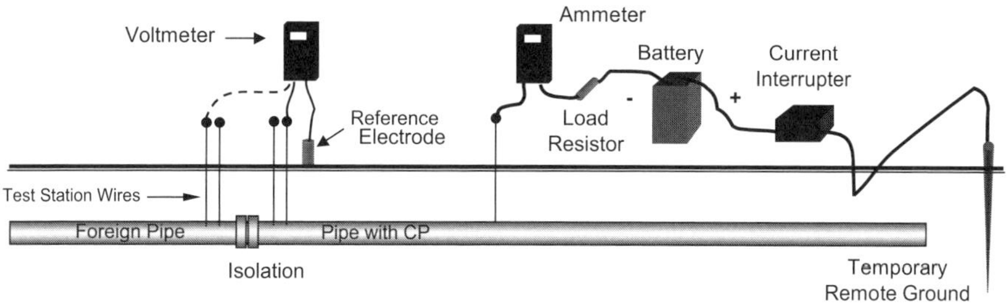

Figure 5.5 Interrupted potential test to determine isolation effectiveness with supplemental interrupted current is shown.

5.2.3 The amount of current being interrupted in Section 5.2.1 or 5.2.2 must cause a change in the structure-to-electrolyte potential greater than 100 mV at the isolation.

5.2.4 With the *reference electrode at the same location for all measurements*, record the ON and OFF structure-to-electrolyte potentials on both the pipe with CP (or structure) and the foreign pipe (foreign structure).

5.2.5 A typical data format is given in Table 5.2.

5.3 Isolation Checker

The objective of this test is to determine if a particular isolation feature is shorted, especially where more than one isolating feature exists at a facility. If the isolating feature is an isolating flange type with double-isolating washers, then a shorted flange bolt can also be identified. The isolation checker operates at a frequency that attenuates quickly along the pipe. The signal

Table 5.2 Typical Field Notes for Interrupted Current Source

	Structure-to-Electrolyte Potential (−mV)				Structure-to-Foreign Potential (−mV)		
	Structure with CP		Foreign Structure				
Location	On	Off	On	Off	On	Off	Remarks

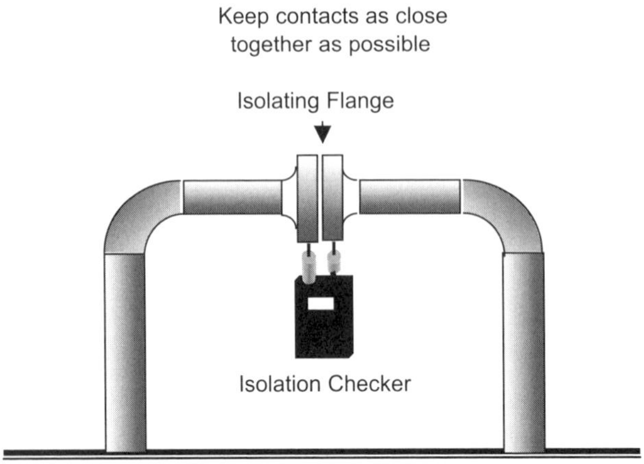

Figure 5.6 Testing an isolator with an isolation checker is shown.

recorded by the instrument is essentially that between the two instrument contacts, that is, only through the isolation being tested. There must be good contact between the instrument probe and the pin; otherwise, the instrument will see the isolation as being effective, regardless of its condition.

5.3.1 Select two contact points on either side of the isolator as close together as possible and clean the surface of the metal at those locations. If the isolating feature uses bolts, *do not* contact each end of the same bolt as that only records the continuity through the bolt.

5.3.2 Turn the isolation checker on and calibrate the instrument. With the probes apart, the meter readout should be at full scale. With the two probes in contact, the meter will deflect toward zero. If the instrument has an audible alarm, it will change from a slow to a rapid beep, ending in a continuous sound.

5.3.3 With the instrument still on, place the instrument probes on the preselected contact points (Section 5.3.1), as shown in Figure 5.6.

5.3.4 Watch the instrument meter to see if it goes toward zero and listen to the alarm, if the instrument is so equipped.

5.3.5 If the instrument meter goes toward zero and the alarm (if equipped) increases in frequency or becomes steady, record the isolating feature as being shorted.

5.3.6 If there is no change in either the instrument meter or the sound, then confirm that both probes are making good contact by leaving one in the same position and moving the other to the same side of the isolation. If the first contact is good, a shorted indication will be observed. Repeat the process on each side of the flange until there is confidence that both probes are in good contact.

5.3.7 An intermittent shorted indication by the instrument may mean that there is a poor probe contact. Confirm that the probes are making good contact by following the procedure described in Section 5.3.6.

5.3.8 With confidence in the probe contact and no instrument meter response or change in the alarm, record the isolating feature as being effective.

5.3.9 If the shorted isolating feature is an isolating flange type with double-isolating washers, conduct the following test on each flange bolt to determine which, if any, are shorting.

- 5.3.9.1 Contact the flange face with one probe and one end of a bolt with the other probe.
- 5.3.9.2 The instrument meter will go downscale toward zero, and the alarm will increase, if the isolated bolt is shorted to the flange.
- 5.3.9.3 Repeat the procedure in Section 5.3.9.1 for all of the bolts.
- 5.3.9.4 Note that a double-washer bolt contacting only one side of the flange will not cause the isolating flange to be shorted. This is the same as an isolating flange with only one set of isolating washers.

5.4 Current Tests

Isolating features can be tested where the current in a structure can be measured on each side of the isolator and at intervals.

5.4.1 It is recommended to install an interrupter in an influencing CP source to give data at two current outputs.

5.4.2 If the isolator is aboveground, measure the current with a clamp-on ammeter on each side of the isolator and toward the current source. If the current is the same and is in the same

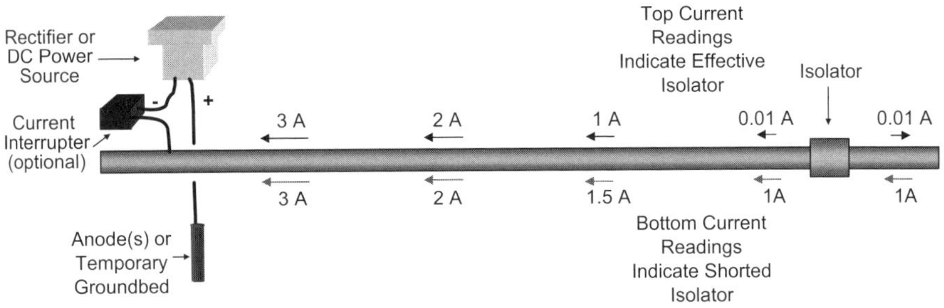

Figure 5.7 Pipeline current used to determine isolation effectiveness is shown.

direction on each side of the isolator, record the isolator as being shorted; otherwise, record it as isolated.

5.4.3 If the isolator is underground, measure the current using a pipeline section as a current span.[4]

5.4.4 If the current measurements are inconclusive, increase the current by establishing a temporary CP installation, as shown in Figure 5.7.

5.4.5 Repeat and record the current measurements in the vicinity of the isolator. Two sets of current values are shown in Figure 5.7 to illustrate an effective isolator (top) and a shorted isolator (bottom).

5.5 Tracing a Sonic Signal

The object of this test is to apply a sonic signal between the two sides of the isolating features and trace that signal through the one that is shorted. This can be an efficient means of locating a shorted isolation where there are multiple parallel isolating features such as at a pipeline header or between utilities.

5.5.1 Parallel Piping

5.5.1.1 Connect a transmitter across one isolator in a group of isolating features or from one side to an electrical ground, as shown in Figure 5.8. Running the transmitter wires

---

[4] *See* Cathodic Protection Procedure No. 3: Direct Current Measurements.

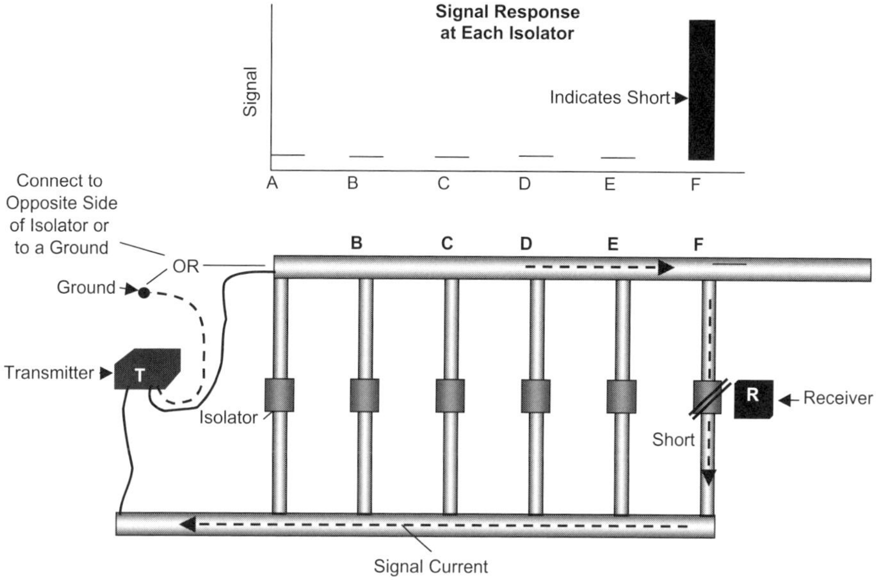

Figure 5.8 Tracing shorted isolation with a sonic signal is shown.

away from the piping to be tested as a signal from the wires will interfere with the signal on the pipe.

5.5.1.2 Trace the signal along each of the piping runs with the receiver, as shown in Figure 5.8. The signal may go through a metallic bypass rather than through an isolator. Record the isolating feature(s) and or metallic bypass with the strongest signal as being shorted, as illustrated by the graph at the top of Figure 5.9.

5.5.1.3 If in doubt, hold the receiver at an isolating feature and intentionally short it with either a jumper wire or a metallic object, as shown in Figure 5.9.

5.5.1.4 If the signal does not change, record the isolation as being shorted (Figure 5.9, response C). If the signal increases in strength, then record the isolation as being effective (Figure 5.9, responses A and B). The bar graph at the top of Figure 5.9 illustrates the signal responses with and without a temporary bond across each isolator in turn.

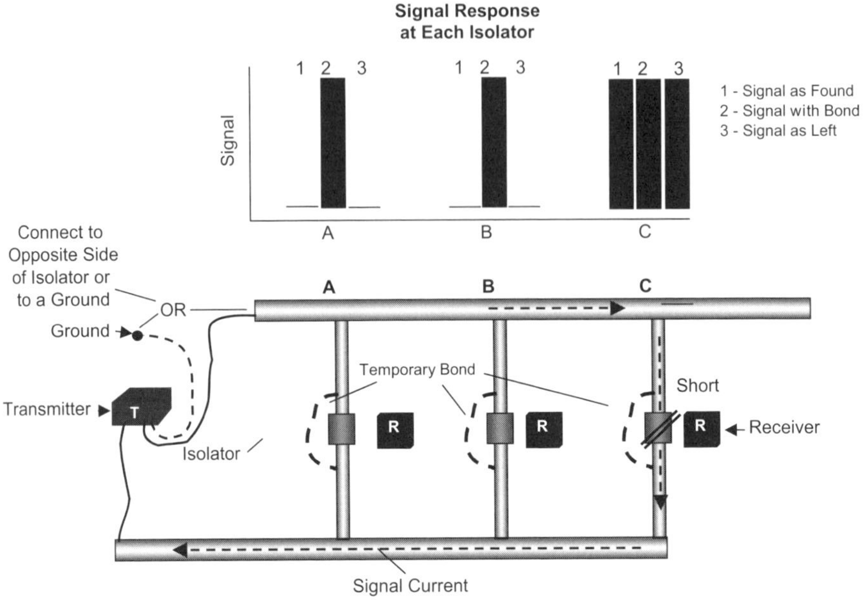

Figure 5.9 Confirming a short by the signal response to a temporary bond is shown.

5.5.2 Utility Piping

5.5.2.1 Utility piping must often be isolated from other utilities by an isolating feature at each service. Structure-to-electrolyte potential methods will identify the fact that the two are shorted but will not define the actual shorted isolation. Similar structure-to-electrolyte potential changes will be noted across an effective isolation that has a shorted service nearby, as with a shorted service isolator.

5.5.2.2 Install a transmitter across a service isolation that is known to be effective (see Figure 5.10).

5.5.2.3 With the receiver, trace the strongest signal along the utility main each way from this service. In Figure 5.10, the strongest signal at D is toward E.

5.5.2.4 Continue to trace the strongest signal along the pipeline. In Figure 5.10, the signal will drop going from E to F but will continue strong toward K. At K, the direction of the strongest signal is now toward G, but there will be a significant drop in signal at H, indicating that the area of

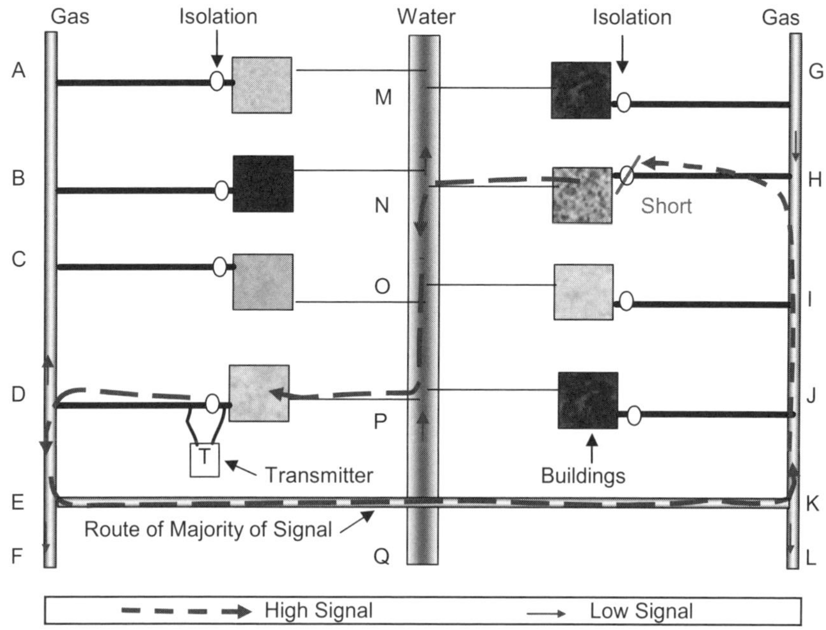

Figure 5.10 Tracing a sonic signal in a utility distribution system is shown.

the shorted service has been found. Confirm the shorted service by tracing the signal through the faulty isolation or through a bypass, whichever case is found. In Figure 5.10, this would be from H to N. The return signal is along the water line from N to P; however, this is seldom traced as the short is discovered by this time.

5.5.2.5 To test large utility systems, receivers and coils mounted in a vehicle will speed the preceding procedure. A handheld receiver is still required to test the shorted service.

5.5.2.6 Note that this is a conductive signal that has a return path through the shorted utility and a parallel path through the earth, or even on isolated utilities. In Figure 5.10, a low signal is shown by small arrows from D to C, E to F, G to H, and K to L to illustrate the return current from pickup through the earth.

5.5.2.7 If the signal current pickup through the earth is greater than that in the pipeline through the shorted service, there will be no definitive response toward the shorted service.

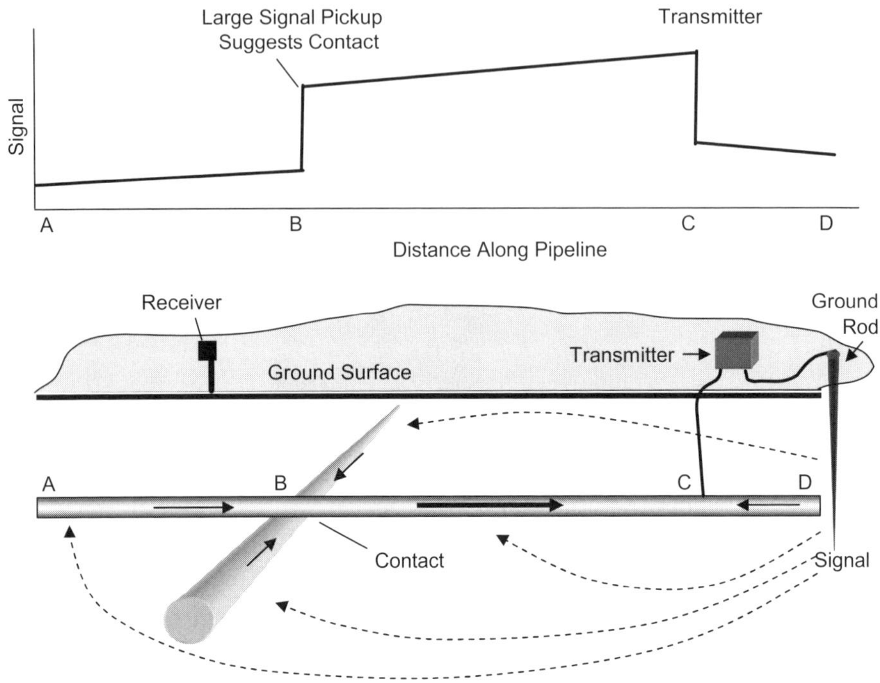

Figure 5.11 Tracing a sonic signal to an underground contact is shown. The graph indicates signal strength along the pipeline from A to D.

Move the transmitter until there is a strong signal trace to the shorted service.

5.5.2.8 Record the service with the strongest signal as being shorted and whether it is the isolator or a bypass.

5.6 Underground Contacts

5.6.1 ON/OFF Potentials

5.6.1.1 Conduct the test similarly to that described in Section 5.2.

5.6.1.2 Use wire of sufficient length to contact both structures suspected of being in contact, while leaving the reference electrode in the same location.

5.6.1.3 Another test will likely be necessary to determine the actual point of contact between the structures, such as that described in Section 5.6.2.

5.6.2 Tracing a Sonic Signal to an Underground Contact

5.6.2.1 Install a conductive transmitter using a sonic signal on the pipeline that is suspected of having an underground contact.

5.6.2.2 Note the strength of the signal on either side of the transmitter and continue in the direction of the strongest signal. In Figure 5.11, the signal strength is shown by the graph on top; thus, the survey would continue from C toward B. At B, a sudden drop in signal will be observed. Confirmation of the contact can be gained by tracing the signal in each direction along the foreign line.

5.6.3 Current Mapping

5.6.3.1 If using a current mapper type of instrument, install the transmitter on the pipeline suspected of having a contact (see Figure 5.12).

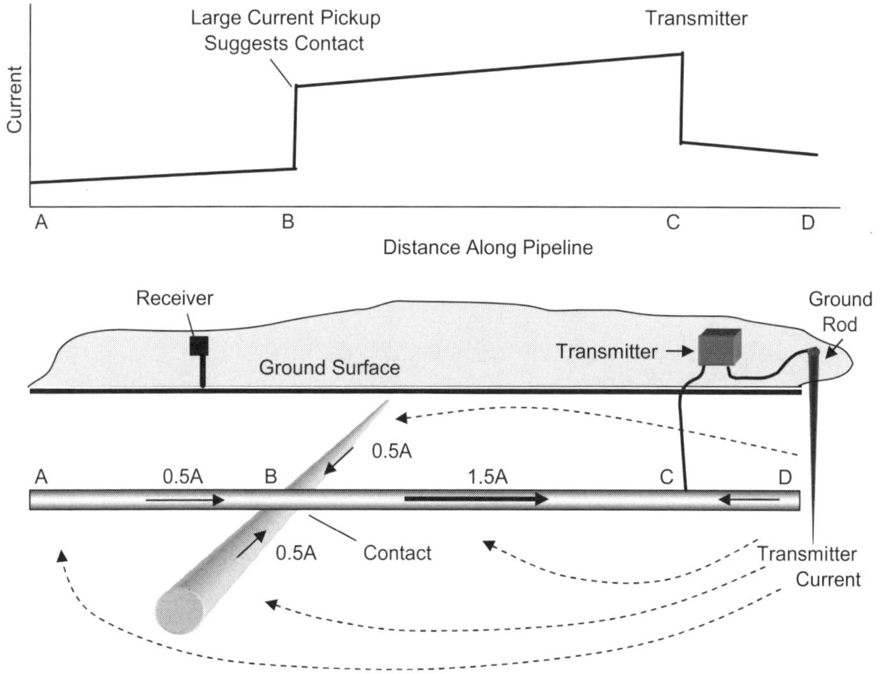

Figure 5.12 Locating an underground contact by pipeline current measurements is shown.

5.6.3.2 Measure the current on either side of the transmitter and proceed in the direction of the higher return current. In Figure 5.12, the higher current is coming from B, rather than D.

5.6.3.3 Measure the current at points along the pipeline to develop a current profile, as seen at the top of Figure 5.12.

5.6.3.4 If there is a significant change in current between any two points, take current measurements at closer intervals in between. The point where a change in pipeline current occurs should be suspected of having an underground contact to a foreign structure. This is illustrated in the profile and in the current values in the profile at the top of Figure 5.12.

5.6.3.5 Note that there will be a normal current pickup along the line, and an indication of a current pickup may also mean that there is a large coating holiday at that location.

5.6.3.6 Move the transmitter to the opposite side of the short and repeat the same test.

5.6.3.7 Plot the results similarly to that shown in Figure 5.13.

5.6.3.8 Where a DC current can be measured by calibrated current spans along the pipe or where a current coil can be

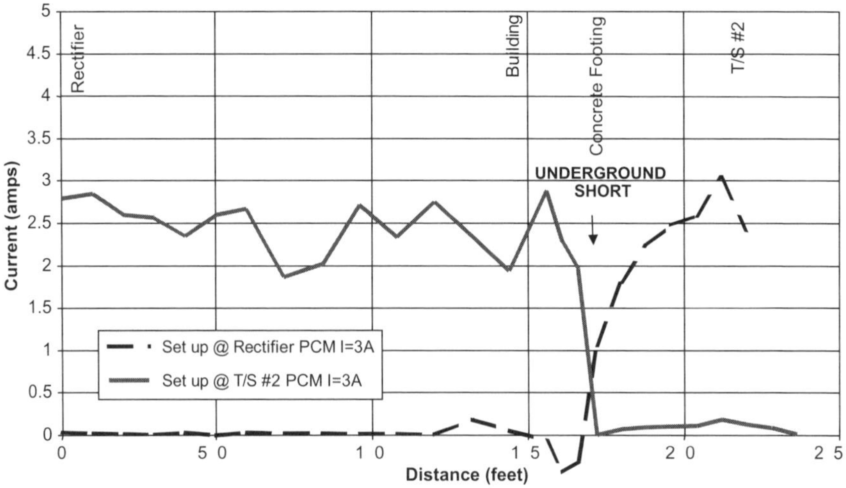

Figure 5.13 Printout from the current mapper surveyed in two directions, confirming the location of the short, is shown.

placed around the pipeline at intervals, a similar procedure is followed.

5.6.3.8.1 Install a current interrupter in a DC power source and measure the current at increments along the pipeline.

5.6.3.8.2 The point of significant current pickup is indicative of an underground contact, although a large coating holiday may also be responsible.

5.7 Isolator Resistance

An isolating feature can be bench tested prior to installation using a resistance method. Following installation, a resistance method may be used, provided certain precautions are followed. These are necessary as much of the test current can use the soil or water parallel path around the isolating feature. In addition, an effective isolation will have a DC voltage existing across the isolating feature that will either aid or oppose the batteries used in the test. For this reason, low-voltage ohmmeters are *not* to be used for testing of installed isolating features.

5.7.1 Bench Testing Isolation on a Nonmetallic Surface without Product Inside

5.7.1.1 If using a high-voltage resistance meter (resistivity meter) with current (C1 and C2) and potential (P1 and P2) terminals, connect the C1 and P1 terminals to one side and the P2 and C2 terminals to the other side of the isolator, as shown in Figure 5.14. The P1 and C1 terminals could be bonded, and, separately, the P2 and C2 terminals could be bonded at the instrument; however, the resistance of the test wires between the terminals and the isolator is added to the measurement.

5.7.1.2 Record the resistance value with the instrument multiplier applied. A low-resistance value (similar to the resistance of the meter leads) indicates a shorted or faulty isolator. The manufacturer of the isolator should specify the expected resistance.

5.7.1.3 If using a low-resistance ohmmeter, connect the leads together to measure the resistance of the test wires.

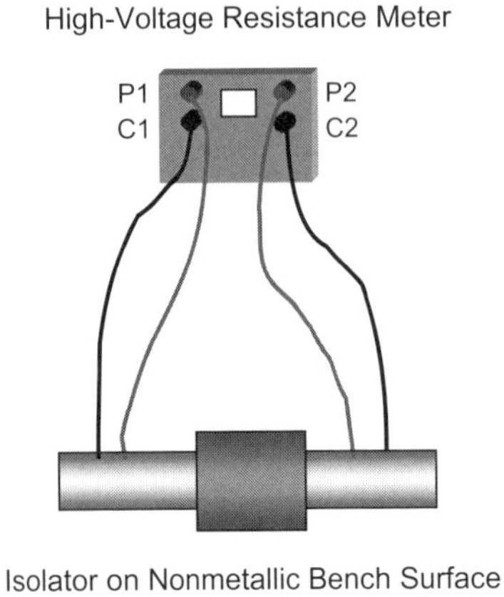

Figure 5.14 Bench testing an isolator with a high-voltage resistance meter on a nonmetallic surface is shown.

5.7.1.4 Connect one test wire to each side of the isolator, as shown in Figure 5.15, making certain that the meter lead contact resistance is minimal, and record the resistance.

5.7.1.5 With a low-voltage ohmmeter, an OL, or infinity, indication on the readout is recorded as an effective isolator, whereas a low-resistance value is recorded as a faulty or shorted isolator.

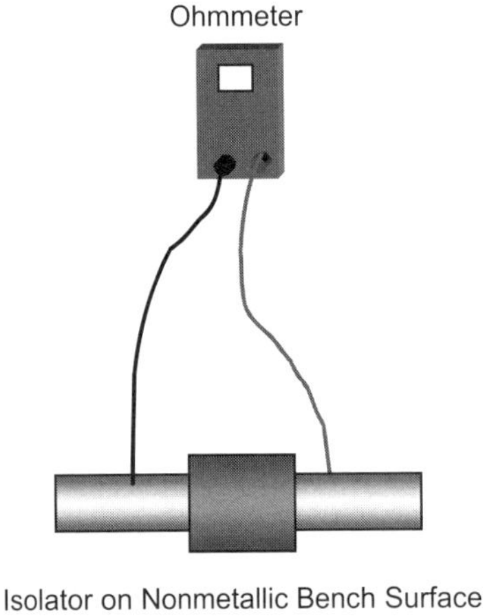

Figure 5.15 Bench testing an isolator with a low-voltage ohmmeter on a nonmetallic surface is shown.

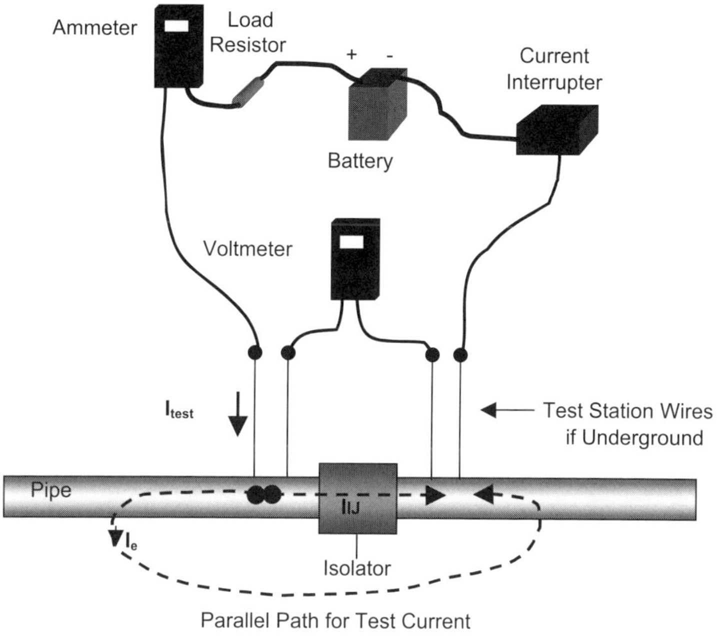

Figure 5.16 Test to determine the resistance of an installed isolator is shown. Note the parallel path for test current through the earth.

5.7.2 Resistance Testing Isolator After Installation

This method of testing an installed isolator is interpretive and should only be conducted and assessed by one very familiar with the expected resistances of the components of the structure.

5.7.2.1 Connect a DC voltmeter across the isolator and measure the DC voltage, noting polarity. A voltage measurement in itself may indicate that the isolator is effective (see Section 5.1). The wires used for the voltage connection are not to be used to carry current.

5.7.2.2 Connect an ammeter, load resistor, battery, and current interrupter in series from one side to the opposite side of the isolator, as shown in Figure 5.16, using wires other than those used for the voltmeter. The control resistor is used to adjust the current but is also necessary because a

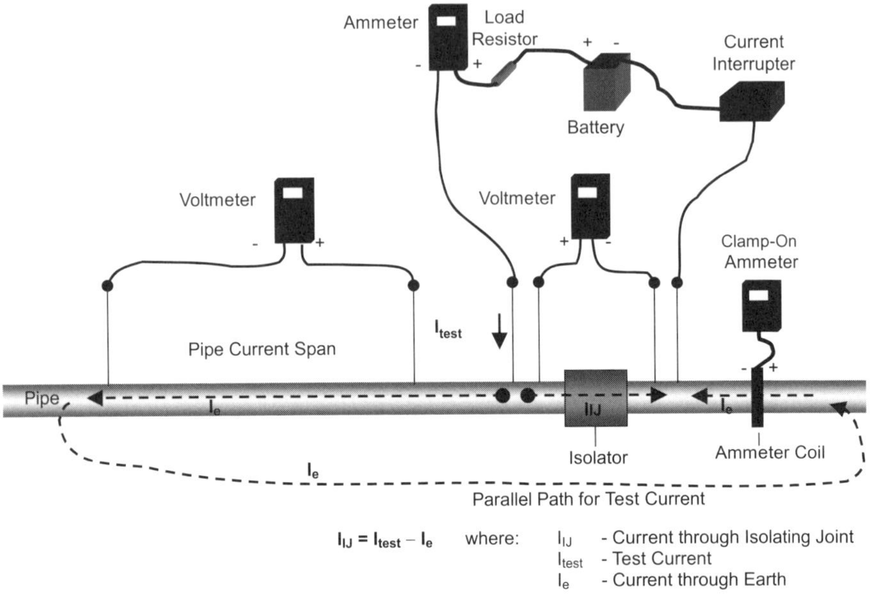

Figure 5.17 Test to determine actual resistance of isolating joint using DC current is shown. Current through the isolating joint can be calculated knowing current through the earth.

shorted isolator will short the battery and the resulting high current will damage the ammeter and current interrupter. A power supply with a controlled output can be substituted for the battery and load resistor.

5.7.2.3 Adjust the test current (*I* test) with the control resistor to yield a significant increase in voltage across the isolator, and record the value and polarity, that is, the direction of current.

5.7.2.4 Measure the DC voltage with the current ON and OFF and record the value and polarity. Note that the polarity may change, depending on the original value, polarity, and the direction of the test current.

5.7.2.5 The test current divides into the current through the isolator and the current through a parallel path through the soil. The resistance of a structure to earth can be very low, such as less than 0.1 Ω, and therefore an effective isolator may still record a low resistance value. Complete the analysis in Section 6.0 before coming to a conclusion on the status of the isolator.

5.7.2.6 If possible, measure and record the current in the pipe going away from the isolator on either side, as shown in Figure 5.17. This current, which is shown as "Ie" in the figure, may be measured either by a pipeline current span or by a clamp-on ammeter.

6.0 ANALYSIS

6.1 Direct Potential Method

6.1.1 A potential difference between the pipe or structure with CP and the foreign pipe or structure that is greater than 100 mV_{DC} is indicative of an effective isolation. A potential difference that is less than this amount does not necessarily mean that the isolation is shorted; however, the results are not considered definitive, and another test is required.

6.1.2 Sample readings are given in Table 6.1.

6.1.3 Case 1 in Table 6.1 suggests that the isolation is shorted but that the potentials on both sides may coincidentally be the same, and

Table 6.1 Sample Field Notes and Analysis for Direct Potential Method

Location	Structure-to-Electrolyte Potential (mV)		Structure-to-Foreign Potential (mV)	Remarks
	Structure	Foreign		
Case 1	870	870	0	Indicates isolation is shorted but must confirm with more tests
Case 2	890	870	20	Isolation status unconfirmed, more tests needed
Case 3	1020	870	150	Indicates isolation is isolated

thus further testing should be done as confirmation. Case 2 shows a small difference between the potentials on either side, but this is not a large enough difference to confirm the status. The difference in potentials in case 3 is greater than 100 mV_{DC}, and therefore the isolation can be considered effective.

6.1.4 The actual location of the short must be determined either by an isolation checker or by tracing a sonic signal.

6.2 Interrupted Current Source Method

6.2.1 If the ON potentials on both the pipe or structure and the foreign structure are the same *and* the OFF potentials are also the same, then the isolator is considered shorted. If either the OFF or ON potentials are different by more than 100 mV_{DC}, the isolator is considered to be effective. If potentials are different but less than 100 mV_{DC}, the status is considered to be unconfirmed, and additional current should be applied or another type of test performed.

6.2.2 Two examples of data are given in Table 6.2, with an analysis. In case 1, the ON potentials of both the pipe and the foreign structure are the same, and the OFF potentials are the same for each also, thus it is shorted. Even though the ON potentials in case 2 are the same, the OFF potentials are 100 mV different, and thus the isolator is considered to be effective. It is not required that both the ON and the OFF potentials between the pipe and the foreign structure be different for an effective isolator.

Table 6.2 Sample Field Notes and Analysis for Interrupted Current Method

	Structure-to-Electrolyte Potential (mV)				Pipe-to-Foreign Potential (mV)		
	Pipe		Foreign				
Location	On	Off	On	Off	On	Off	Remarks
Case 1	−970	−870	−970	−870	0	0	Isolator Shorted
Case 2	−970	−870	−970	−970	0	100	Isolator Effective

6.3 Isolation Checker

6.3.1 The isolation checker is a direct readout that indicates whether the isolator is effective or shorted.

6.4 Current Tests

6.4.1 An isolator that will not allow pipeline current to pass is considered isolated. Pipeline current measured immediately on either side of an isolator must be zero for the isolator to be effective. Near the isolator, a small current going away from it on each side indicates an effective isolator. As this test may not be conclusive, a second measurement at a different test current value or another test should be conducted as confirmation. An example of an effective isolator is shown by the top values in black in Figure 6.1.

6.4.2 Current will pass through a shorted isolator. A similar value of current measured on each side of the isolator and in the same direction suggests that it is shorted. However, a current may pass by an effective isolation by going around through the soil or by going through the product inside the pipe (stray current interference). If the isolator is aboveground, then current measurements on either side on aboveground piping will not record the current in the soil. Otherwise, current bypassing

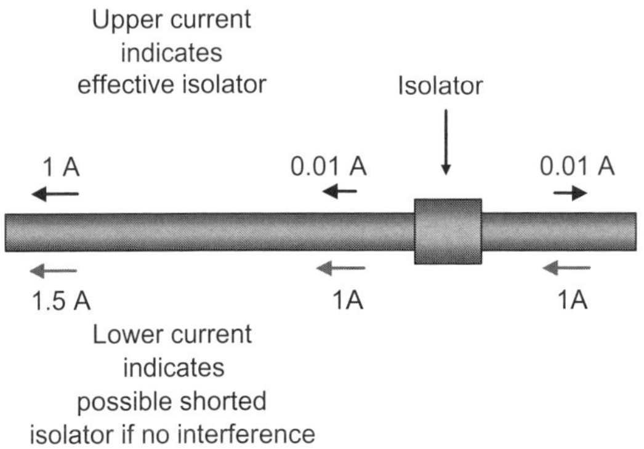

Figure 6.1 Isolator test using pipeline current is shown.

through the soil can be confirmed by an electropositive shift in the structure-to-electrolyte potentials[(5)] on one side of the isolator and an electronegative shift in potentials on the other side when this current is interrupted. Unfortunately, this does not apply to a current bypassing through the product inside the pipe; in this case, however, a low-resistivity product inside should arouse suspicion. An example of a shorted isolation based on current measurements without stray current interference is shown by the lower values in Figure 6.1.

6.5 Tracing a Sonic Signal

6.5.1 Parallel Piping

6.5.1.1 An effective isolator will block the signal from a conductive transmitter. When the signal dies as the isolator is approached, the isolator is effective. This can be confirmed by temporarily bonding or shorting the isolator, thus allowing the signal to pass. If the signal increases when it is temporarily bonded or shorted, the isolator is effective (Figure 5.10).

6.5.1.2 The signal from a conductive transmitter will continue through a shorted isolator when there is no change in signal. As the receiver goes past the isolator, it is shorted. This can be confirmed by temporarily bonding or shorting the isolator, thus ensuring that the signal passes. If the signal does not change with this temporary bond or short, the isolator was already shorted (Figure 5.10).

6.5.2 Utility Piping

6.5.2.1 Utility piping is similar to parallel piping, except that the isolators are varied and widely distributed.[7.2] The signal is then used as a trace to quickly find the shorted isolator.

6.5.2.2 A large amount of the conductive transmitter signal current will travel along the pipeline through the shorted isolator to return through the other utility. A portion of

[5] *See* Cathodic Protection Procedure No. 8: Direct Current Stray Current Interference.

this signal will also be picked up by the piping through the earth. The latter signal will be seen on all piping probably in proportion to the length of piping beyond the point at which the signal is being recorded.

6.5.2.3 If the signal current pickup onto the pipe (structure) is larger than that through the isolator, then the signal trace to the shorted service will be indistinguishable, and the transmitter should be moved until there is a distinct signal to the shorted isolator.

6.5.2.4 The signal from a conductive transmitter will continue through a shorted service; therefore, when there is no change in signal as the receiver goes past the isolator, it is shorted. This can be confirmed by temporarily bonding or shorting the isolator, thus ensuring that the signal passes. If the signal does not change with this temporary bond or short, the isolator was already shorted.

6.6 Underground Contacts

6.6.1 Using ON/OFF Potentials to Determine an Underground Short

6.6.1.1 If the ON potentials on both structures are the same *and* the OFF potentials are also the same, then the structures are considered to be electrically continuous or in electrical contact. If the OFF potentials or the ON potentials measured on each side of the isolation differ by more than 100 mV_{DC}, the structures are considered to be isolated. If potentials differ but are less than 100 mV_{DC}, the status is considered to be unconfirmed, and additional current must be applied or another type of test performed.

6.6.1.2 If a contact is indicated, another test should be completed, such as tracing a sonic signal or current mapping, as confirmation and to locate the point of contact.

6.6.2 Sonic Signal Trace to an Underground Short

6.6.2.1 A sonic signal from a conductive transmitter will be picked up by all metallic structures that are connected to the pipeline, whether intentionally or accidentally.

6.6.2.2 The signal trace near the receiver is a combination of that being picked up by the pipeline and that being picked up by other structures that are in contact. Beyond the contact, the signal drops to just that being picked up by the pipeline.

6.6.2.3 If the contact is at a point where the signal has attenuated to near zero, the change in signal at the contact may not be discernable. In this case, the transmitter should be moved until the overall pipeline section has been tested with a strong signal.

6.6.2.4 A significant change in signal strength will be detected at the contact. The signal can also be followed along the foreign structure as confirmation.

6.6.3 Current Mapping

6.6.3.1 Under uniform coating conditions, the pipeline current will steadily increase toward the DC power source. A sudden change over a short period of time indicates that there is an abnormal amount of bare metal exposed to the soil. This may be due to missing coating, but a large increase in current going toward the current source is most likely a contact to a foreign structure.

6.7 Isolator Resistance

6.7.1 Bench Testing Isolation on a Nonmetallic Surface Without Product Inside

6.7.1.1 A measurement that exceeds the capacity of the high-voltage resistance meter indicates an effective isolator.

6.7.1.2 A reading that is less than 10,000 Ω is considered to be a faulty isolator (although normally, a shorted isolator will have a resistance less than 10 Ω).

6.7.1.3 A measurement that reads OL on a digital ohmmeter indicates an effective isolator.

6.7.2 Resistance Testing Isolator After Installation

6.7.2.1 With current applied across an isolator, its resistance, including the parallel path through the earth and the parallel path through the product, if it is a liquid, can be calculated by Equation (6.1):

$$R_t = \frac{\Delta V}{I_{\text{test}}}, \tag{6.1}$$

where

- R_t total resistance of isolator and parallel earth path (ohms)
- ΔV (ON – OFF) DC voltage across isolator (volts)
- I_{test} total current applied during the test (amperes)

6.7.2.2 Before deciding on the effectiveness of an isolator by this method, the resistance to earth of the structure should first be known, as this is part of the parallel circuit being measured. With large structures, the resistance to earth may be 0.1 Ω, and therefore a resistance of this order could still indicate an effective isolator. A resistance of 0.1 Ω on a small system may indicate a shorted isolator. This is further complicated by the parallel path through the product, especially if it is a low-resistivity liquid. A total resistance (R_t) less than the structure-to-earth resistance indicates a shorted isolator, whereas a total resistance much higher indicates an effective isolator.

6.7.2.3 If the earth current can be determined by measuring the pipeline current, then the actual resistance of the isolator and the parallel path through the product, if it is a liquid, can be calculated by Equation (6.2):

$$R_i = \frac{\Delta V}{(I_{\text{test}} - I_e)}, \tag{6.2}$$

where

- R_i resistance of isolator and parallel product path (ohms)
- ΔV (ON – OFF) DC voltage across isolator (volts)
- I_{test} total current applied during the test (amperes)
- I_e current through the earth or pipeline current (amperes)

6.7.2.4 The accuracy of the isolator resistance depends in part on the accuracy of the earth current (I_e) measurement as that can represent a large part of the test current (I_{test}) and indicate if there is a low-resistance parallel path through the product.

6.7.2.5 With a dry or high-resistivity liquid product, an isolator resistance, as calculated by Equation (6.2), of greater than 1,000 Ω can be expected if it is effective. A low resistance indicates that it is shorted.

7.0 REFERENCES

7.1 NACE Standard SP0169-2007, "Control of External Corrosion on Underground or Submerged Metallic Piping Systems" (Houston, TX: NACE International, 2007).

7.2 NACE Standard SP0286-2007, "Electrical Isolation of Cathodically Protected Pipelines" (Houston, TX: NACE International, 2007).

7.3 J.E. Wright, *Practical Corrosion Control Methods for Gas Utility Piping,* 2nd ed. (Houston, TX: NACE, Item # 37542, 1995), pp. 41–51.

7.4 NACE Standard SP0177-2007, "Mitigation of Alternating Current and Lightning Effects on Metallic Structures and Corrosion Control Systems" (Houston, TX: NACE International, 2007).

APPENDIX A
DECISION TREE FOR ISOLATION TESTS

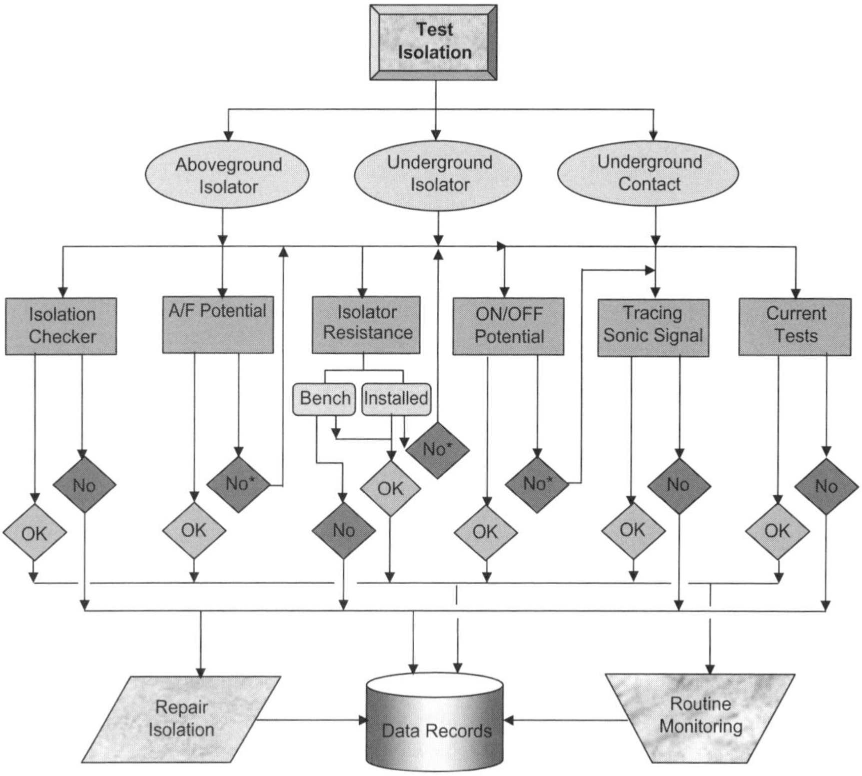

* Confirm with another test

CATHODIC PROTECTION PROCEDURE NO. 10

Road and Railroad Cased Crossings (Basic)

10

1.0 INTRODUCTION

The purpose of this procedure is to outline a procedure to test the electrical isolation of road or railroad cased crossings.

When the annular space between the casing and the pipe is dry, cathodic protection (CP) current will not access the inside of the carrier pipe; however, the only corrosion will then be due to oxidation from the humidity inside.

A casing may be coupled to the carrier pipe by a metal contract (electronic), by water or mud in the annular space between the pipe and casing (electrolytic), or by a combination of the two. When an electrolytic couple exists between the pipeline and the casing with the casing isolated, CP current will pass through the casing to provide protection to the carrier pipeline inside. This is sometimes called an electrolytic short, but in fact, it is not an electrical short circuit, and thus in this procedure, it will be called an "electrolytic couple."

CP is ineffective when a low-resistance metallic contact exists between the carrier pipe and the casing (called "shorts" or "electronic shorts") as the majority of the CP current will take the metallic path, rather than the electrolytic path, through the annular space to the pipeline.

Even though a structure-to-electrolyte potential measured at the ground surface may meet a criterion for CP, this measurement only reflects the condition on the outside of a shorted casing and/or the pipeline to each side. The pipe inside a shorted casing will be shielded from CP; therefore, the pipe is not protected against corrosion. The majority of CP current will preferentially return to the source via the casing wall to the point of contact, and then to the pipe itself, leaving little current to be impressed onto the pipe inside the casing.

In general, there are six methods to test for a couple between the carrier pipe and the casing, including the following:

- Steady potentials
- OFF/ON potentials
- Forced depolarization
- Potential attenuation
- Current attenuation
- Annulus resistance

The first three are covered in this procedure, in addition to a method to predict the location of the electronic couple.

Potential attenuation is determined by measurements taken at test stations or during a close interval survey, which can be used as a qualitative test to predict that a casing is shorted by a more rapid attenuation of potentials at the casing. This is covered in other procedures.[(1)] Current attenuation is determined by consecutive measurements of pipeline current, where a large current pickup at the casing indicates a shorted casing. The facilities required to measure current are seldom available, and therefore this procedure has not been covered. The measurement of the annulus resistance requires expert analysis as a parallel circuit of low resistance exists through the soil, and for this reason, this procedure is also not covered here. NACE SP0200[7.1] provides a description of casing tests. As Barlo and Lopez concluded, more than one test should be used to assess the isolation of a cased pipeline.[7.2]

With the exception of some tests related to the rectifiers, CP Testers, CP Technicians, CP Technologists, and CP Specialists or equivalent are qualified to complete the field tests. See "Cathodic Protection Procedure No. 1: Rectifier Adjustment, Inspection, and Basic Troubleshooting" for qualifications to work on rectifiers. CP Technologists and CP Specialists or equivalent should complete the analysis.

---

1 *See* Cathodic Protection Procedure No. 2: Structure-to-Electrolyte Potential Measurement, Cathodic Protection Procedure No. 4: Diagnostic Testing (Current Requirement), and Cathodic Protection Procedure No. 7: Close Interval Potential Survey.

A hazard assessment and a job safety analysis are to be completed for each phase of the project, and the appropriate preventative measures are to be taken.[2]

2.0 TOOLS AND EQUIPMENT

- Multimeter capable of measuring 1 mV_{DC} to 40 V_{DC}, complete with leads with insulated probes and clips
- Copper-copper sulfate reference electrode

The following equipment will depend on the test selected:

- Current interrupter
- Direct current (DC) ammeter sized for test current
- Portable DC power supply (6–12 V_{DC})
- If power supply is not regulated, temporary resistance load sized for the current and wattage of the power supply
- Test wires as necessary for the applicable test
- Small hand tools

3.0 SAFETY EQUIPMENT

- Standard safety equipment and clothing, as required by the company's safety manual and regulations
- Lock-out/tag-out kit
- Electrically insulated clips and probe handles for meter leads

4.0 PRECAUTIONS

4.1 An electrolytic couple between the pipe and the casing may at times display similar test results to an electronic short (metallic contact). This is different from an electronic short as CP current is not shielded by the casing and can still continue through the casing to the pipe's surface. In effect, the casing is coupled to the pipe through the resistance of the electrolyte in the annulus. This is a

[2] See the owner's or your company's Safety Manual.

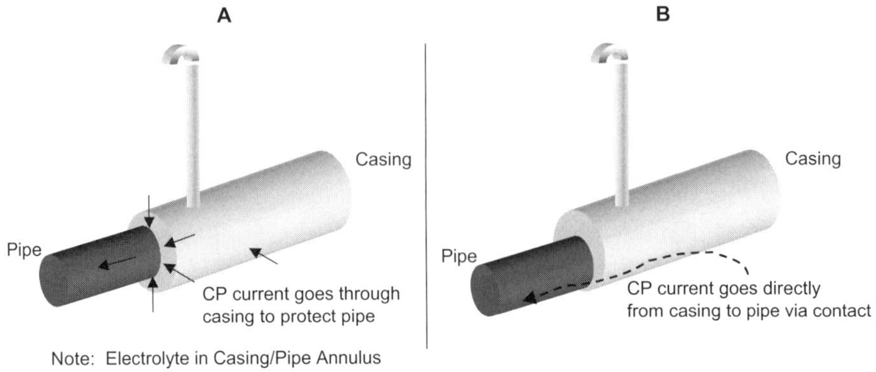

Figure 4.1 (A) Electrolytic couple between pipe and casing and (B) electronic short or metallic short between pipe and casing are shown.

condition that can cause a high current to be impressed onto the poorer coated or bare casing. An electrolytic couple is sometimes called an "electrolytic short." A "short" in this document will only refer to a metallic contact between the pipe and the casing. The two situations are illustrated in Figure 4.1.

4.2 Note that the reference electrode is to remain at the same location for structure-to-electrolyte potentials intended to determine continuity.

4.3 The voltmeter is to be capable of measuring negative values, allowing the negative lead of the voltmeter to be connected to the copper-copper sulfate reference electrode and the positive voltmeter lead to the appropriate test lead and/or casing vent. When the meter test leads are connected in this manner, the voltmeter will display a negative meter polarity, a potential value, and the units in millivolts or volts. If the meter is not capable of measuring a negative reading, the leads may be reversed, but the value must still be recorded as a negative value.

5.0 PROCEDURE

One or more of the following tests can be completed to confirm that the pipeline is electrically isolated from the casing. A flowchart to assist in decision making is shown in Appendix A of this procedure.

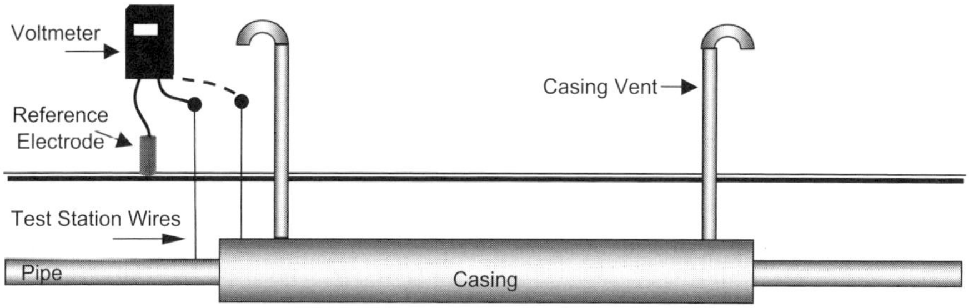

Figure 5.1 Direct potential method to test a casing isolation is shown.

5.1 Direct Pipe-to-Electrolyte Potential Method

5.1.1 Measure a structure-to-electrolyte potential[(3)] on the pipeline and another on the casing *with the reference electrode remaining at the same location*, normally, near the end of the casing (see Figure 5.1). Alternately, measure a potential difference between the pipe and casing directly.

5.1.2 Sample field notes for a direct pipe-to-electrolyte test are given in Table 5.1.

Table 5.1 Typical Field Notes for Direct Potential Method

Location	Pipe-to-Electrolyte Potential (mV)		Pipe-to-Casing Potential (mV)	Remarks
	Pipe	Casing		

5.2 Interrupted Current Source Method

The objective of this method is to change the structure-to-electrolyte potential on the pipeline that will not be the same on the casing if the pipe and casing are isolated.

3 *See* Cathodic Protection Procedure No. 2: Structure-to-Electrolyte Potential Measurement.

Table 5.2 Typical Field Notes for Interrupted Current Source

	Pipe-to-Electrolyte Potential (mV)				Pipe-to-Casing Potential (mV)		
	Pipe		Casing				
Location	On	Off	On	Off	On	Off	Remarks

5.2.1 Interrupt one of the existing CP DC power sources.

5.2.2 Alternately, apply a supplemental interrupted CP current from a temporary DC power source and ground a sufficient distance away so that the casing is not within the voltage gradient of the interrupted anodes. Figure 5.2 illustrates a typical test setup.

5.2.3 The amount of current being interrupted in Section 5.2.1 or 5.2.2 must cause a change in potential greater than 100 mV.

5.2.4 With the *reference electrode at the same location for all measurements*, record the ON and OFF structure-to-electrolyte potentials on both the pipe and the casing.

5.2.5 A typical data format when noting the effects on a structure-to-electrolyte potential from an interrupted current is given in Table 5.2.

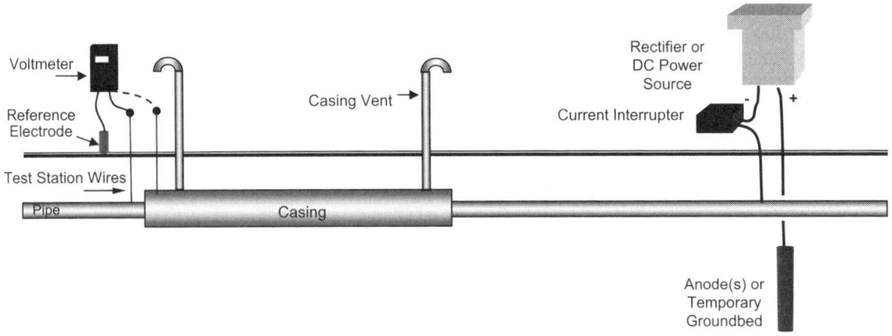

Figure 5.2 Interrupted potential test to determine casing isolation is shown.

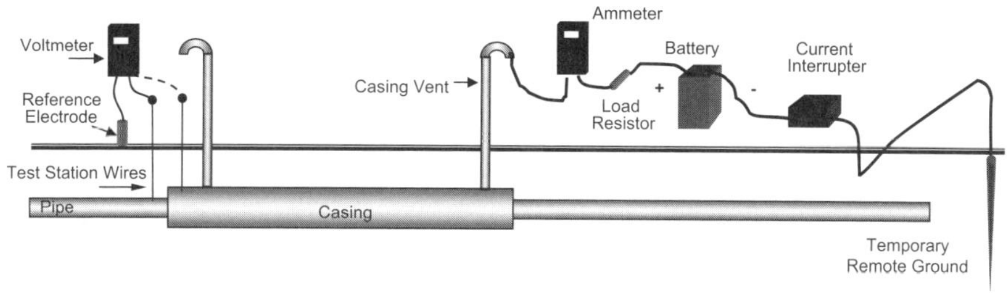

Figure 5.3 Casing depolarization test to determine casing isolation is shown.

5.3 Temporary Casing Depolarization Test

The objective of this test is to depolarize the casing temporarily to separate the structure-to-electrolyte potentials between the pipe and the casing.

5.3.1 Measure the as-found structure-to-electrolyte potential of the pipe and the casing before any additional current is applied.

5.3.2 Connect the positive terminal of a temporary DC power source[4] to the casing and the negative terminal to a temporary ground, with an ammeter, control resistor, and current interrupter inserted in the connecting cables.

5.3.3 Apply a temporary interrupted current between the pipe and the casing, with the casing acting as an anode, as shown in Figure 5.3. The control resistor is used to control the current output; however, it and the battery can be substituted for a controlled DC power source.

5.3.4 Impress a current large enough to cause a depression in the casing-to-electrolyte potential.

5.3.5 Measure the ON and OFF structure-to-electrolyte potentials of the pipe and the casing.

5.3.6 Increase the current by an increment of current that is near the initial current amount and repeat the interrupted potential measurements on the pipe and the casing.

---

[4] *See* Cathodic Protection Procedure No. 1: Rectifier Adjustment, Inspection, and Basic Troubleshooting.

5.3.7 Increase the current by another increment equivalent to the last increment and repeat the interrupted potential measurements on the pipe and the casing.

5.3.8 Further increases in current are optional; however, the first three are mandatory.

5.3.9 Note that an interference condition on the pipe may result from the casing being used temporarily as an anode.

5.4 Predict Location of Pipe-to-Casing Contact

5.4.1 The approximate location of the contact may be predicted by the following test, which is completed in three parts:

- 5.4.1.1 Calibrating the casing
- 5.4.1.2 Impressing a current from one side of the casing and measure the voltage across the casing
- 5.4.1.3 Impressing a current from the opposite side of the casing and measure the voltage across the casing

5.4.2 Calibrate the Casing

- 5.4.2.1 Measure the voltage between the casing vents or one set of test lead wires.
- 5.4.2.2 Connect the DC power source load resistor and ammeter in series from the ends of the casing using a different set of wires than was used to measure the voltage, as shown in Figure 5.4.
- 5.4.2.3 Apply a current that will cause a measurable millivolt drop between the ends of the casing.
- 5.4.2.4 Record the ON and OFF voltage drop between the ends of the casing.
- 5.4.2.5 Record the current and the ON and OFF voltage measurements.
- 5.4.2.6 Repeat the test at a second current value.

5.4.3 Current Between the Pipe and One End of the Casing

- 5.4.3.1 Measure the voltage between the casing vents or between test leads at each end of the casing.
- 5.4.3.2 Connect the DC power source load resistor and ammeter in series from one end of the casing to the pipeline using a different set of wires than was used to measure the voltage, as shown in Figure 5.5.

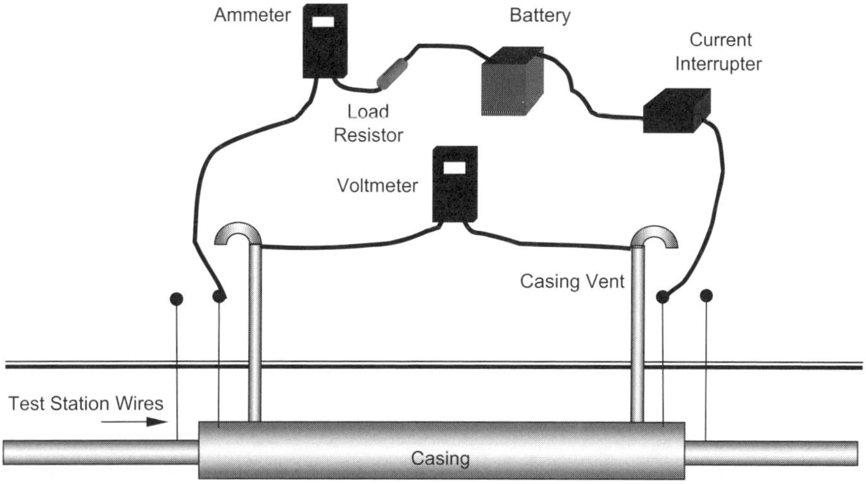

Figure 5.4 Calibrating the casing resistance is shown.

5.4.3.3 Apply a current that is similar to one of the current values that was used in the preceding casing calibration.

5.4.3.4 Record the ON and OFF voltage drop between the ends of the casing.

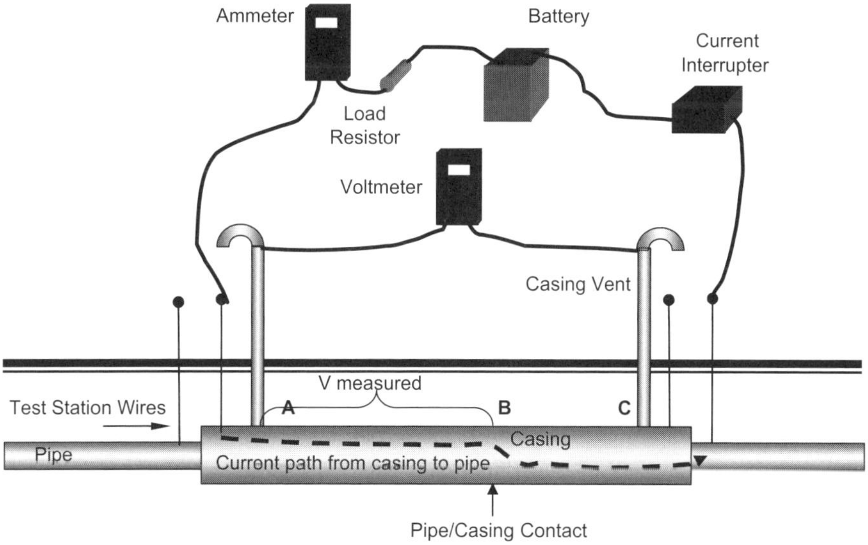

Figure 5.5 Calibrating the casing to the point of contact from the left is shown.

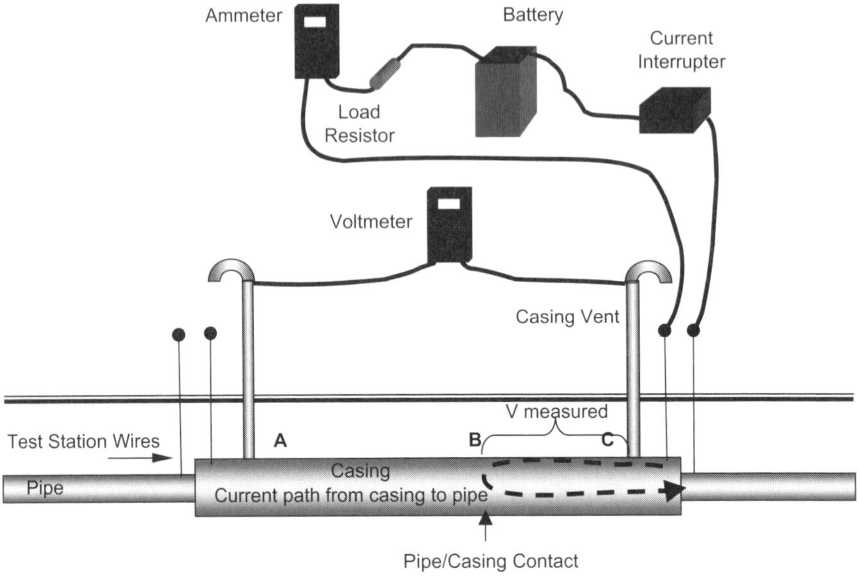

Figure 5.6 Calibrating the casing to the point of contact from the right is shown.

5.4.3.5 Record the current and the ON and OFF voltage measurements.

5.4.3.6 Note that the voltage (casing *IR* drop) is being measured between A and B in Figure 5.5.

5.4.4 Current Between the Pipe and the Opposite End of the Casing

5.4.4.1 Measure the voltage between the casing vents or one set of test lead wires.

5.4.4.2 Connect the DC power source load resistor and ammeter in series from one end of the casing to the pipeline using a different set of wires than was used to measure the voltage, as shown in Figure 5.6.

5.4.4.3 Apply a current that is similar to one of the current values that was used in the preceding casing calibration.

5.4.4.4 Record the ON and OFF voltage drop between the ends of the casing and record the current.

5.4.4.5 Note that the voltage (casing *IR* drop) is being measured between B and C in Figure 5.6.

5.4.4.6 If the same current was used in all three tests, the sum of the voltage drops in the last two tests should be approximately equal to the voltage drop in the first test. If there is a large discrepancy, then the tests should be repeated. If a large discrepancy still exists, there is probably more than one contact between the pipe and the casing.

6.0 ANALYSIS

6.1 Direct Potential Method

6.1.1 A potential difference between the pipe and the casing that is greater than 100 mV is indicative of an isolated casing. A potential difference between the pipe and casing that is less than this amount does not necessarily mean that the casing is shorted; however, the results are not considered definitive, and another test, as described previously, is required.

6.1.2 Sample readings are given in Table 6.1. Case 1 readings suggest that the casing is shorted but the potentials could coincidentally be the same, and thus more testing is needed. Case 2 indicates that the casing is isolated.

6.2 Interrupted Current Source Method

6.2.1 If the ON potentials on both the pipe and casing are the same *and* the OFF potentials are also the same, then the casing is considered shorted. If either the OFF and ON potential on the pipe differs by more than 100 mV_{DC} from the corresponding OFF or ON potential of that on the casing, the pipe is normally considered to be isolated from the casing. If potentials are different but less

Table 6.1 Sample Field Notes and Analysis for Direct Potential Method

	Structure-to-Electrolyte Potential (mV)			
Location	Pipe	Casing	Pipe-to-Casing Potential (mV)	Remarks
Case 1	820	820	0	Indicates casing is shorted
Case 2	920	820	150	Indicates casing is isolated

Table 6.2 Sample Field Notes and Analysis for Interrupted Current Method

	Structure-to-Electrolyte Potential (mV)				Pipe-to-Casing Potential (mV)		
	Pipe		Casing				
Location	On	Off	On	Off	On	Off	Remarks
Case 1	– 870	–770	–870	–770	0	0	Casing Shorted
Case 2	–870	–770	–870	–870	0	100	Casing Isolated

than 100 mV, additional current should be applied to confirm the status or another test performed.

6.2.2 Two examples of data are given in Table 6.2. In case 1, the ON potentials of both the pipe and the casing are the same, and the OFF potentials are the same for each also. Even though the ON potentials in case 2 are the same, the OFF potentials are 100 mV different, and thus the casing is considered isolated. It is not required that both the ON and the OFF potentials between the pipe and the casing be different.

6.3 Casing Depolarization Test

6.3.1 The pipe may become slightly more electropositive or electronegative, but the casing will become significantly more electropositive as the test current is applied if the pipe is isolated from the casing (see Table 6.3, case 1). If a metal-to-metal contact exists, the pipe and the casing potential will become more electropositive by similar amounts (see Table 6.3, case 2). The difference between the pipe and the casing potential will depend on the pipe-to-casing contact resistance.

6.4 Predict Location of Pipe-to-Casing Contact

6.4.1 Calculate the calibration factor of the casing from the results of the first test (Figure 5.4), as follows:

$$F = \frac{\Delta I}{\Delta V}, \quad (6.1)$$

Table 6.3 Typical Field Notes for a Casing Anode Test

Location	Current Applied (A)	Structure-to-Electrolyte Potentials (mV)				Pipe-to-Casing Potential (mV)	Remarks
		Pipe		Casing			
		A/F*	On	A/F*	On		
Case 1		−870		−850		20	
	0.25		−860		−720	140	
	0.50		−850		−580	270	
	0.75		−840		−200	640	
	1.00		−830		+075	905	Casing Isolated
Case 2		−980		−980		0	
	0.25		−920		−920	0	
	0.50		−750		−750	0	
	0.75		−580		−580	0	
	1.00		−400		−400	0	Casing Shorted

*A/F, as found.

where

- F calibration factor (amperes per millivolt)
- ΔI net test current (amperes)
- ΔV (ON – OFF) voltage across casing (millivolts)

6.4.2 From the results of the second test, calculate the percentage distance along the casing from the first casing connection end (Figure 5.6), as follows:

$$X = 100 \times \frac{\Delta \text{mV} \times F}{\Delta I}, \tag{6.2}$$

where

- X distance from casing connection end to contact (percentage of casing length)
- ΔmV (ON – OFF) voltage (millivolts)
- F calibration factor (amperes per millivolt)
- ΔI net test current (amperes)

6.4.3 From the results of the third test, calculate the percentage distance from the opposite end of the casing using Equation (6.2) and compare the results.

6.4.4 The percentage length of the casing from this test plus the percentage distance along the casing from the last test in Section 6.4.3 should be near 100%. If there is a large discrepancy, then there may be more than one contact.

7.0 REFERENCES

7.1 NACE Standard SP0200-2008, "Steel-Cased Pipeline Practices" (Houston, TX: NACE International, 2008).

7.2 T.J. Barlo, A. Lopez, "Cased Crossing Methods," CORROSION/2000, paper no. 00726 (Houston, TX: NACE International, 2000).

APPENDIX A
DECISION TREE FOR TESTING ISOLATION

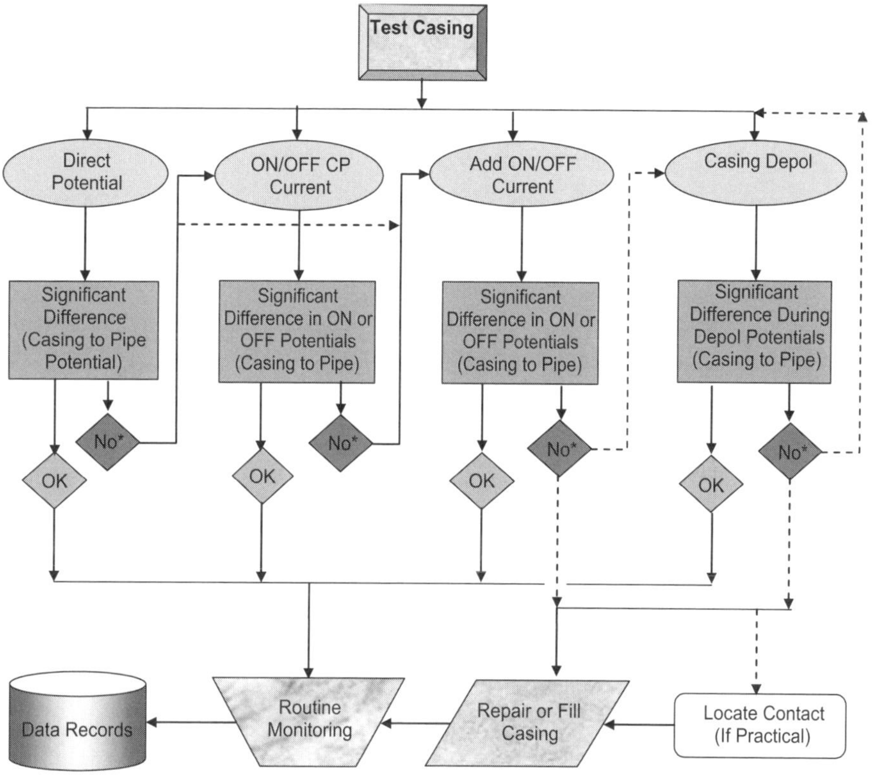

* Confirm with at least two tests

CATHODIC PROTECTION PROCEDURE NO. 11

Alternating Current Voltage Testing 11

1.0 INTRODUCTION

Stray alternating current (AC) on structures is a safety concern, and at high AC current densities, corrosion can also be a concern. Any person coming in contact with a structure that may have a hazardous AC voltage must be aware of these hazards and the safety precautions that need to be taken.

There has been a significant amount of study on the subject since the 1970s.[7.1–7.6] NACE International and the Canadian Standards Association issued a Recommended Practice dealing with this problem.[7.7,7.8]

There are three couplings termed "capacitive," "inductive," and "conductive" (also called resistive), which can cause an AC voltage on a structure such as a pipeline that is in parallel with an electric power line but is not limited to such a case.

Capacitive effects have to be considered on aboveground pipelines and equipment, especially those with poor electrical contact to the ground, such as when pipelines are under construction, and on skids and rubber-tired vehicles. The AC voltage normally increases on a pipeline during welding operations as the length of the aboveground section of pipe is increased.

An inductive coupling is caused by the changing magnetic field from the AC in the power transmission line. An AC voltage will be induced on a pipeline that passes through this changing magnetic field. The AC voltage induced is dependent on the length of the parallel section, the relative spacing to the conductor, or between each phase conductor and the pipeline, the coating effectiveness, the soil resistivity, and the power line load. Once the pipeline and power line are installed, the AC voltages on the pipeline will be essentially permanent but will vary with the power line load and from point to point along the pipeline.

A conductive or resistive coupling is a result of an AC current that collects on a pipeline from a close-proximity power transmission line ground. A large fault current or lightning strike that enters the pipe will not only cause large voltage gradients, but may cause damage to the pipe and/or coating if it enters at a high current density. Research indicated that not only can molten pits occur causing failures, but cracks may also develop around the molten area (assuming that penetration has not taken place).[7.4]

The prediction of AC voltages is complex. In general, the voltage will peak at discontinuities between the pipeline and the power line and attenuate exponentially between the discontinuities to the point that portions of the pipeline may have little or no induced voltage. A discontinuity between the power line and the pipeline may involve the following:

- A change in the transmission line to pipeline separation
- The end or beginning of the parallel exposure
- A power line transposition
- A change in the number of line conductors or pipelines in the common right-of-way

Discontinuities due to changes in the pipeline characteristics are sometimes more difficult to determine because the information is not readily apparent from drawings or file information. Changes in the characteristics can arise from any of the following reasons:

- A change in the coating conductivity
- An extreme change in soil resistivity (somewhat dependent on coating)
- A change in the pipe size or thickness
- An interruption in the pipe continuity (isolating features)

Under fault conditions, a voltage gradient will develop in the area of current discharge to the earth and around the pipeline if that becomes part of the circuit. This, together with the body resistance to earth, the duration of the shock, and the physical condition of the person, will determine the safe

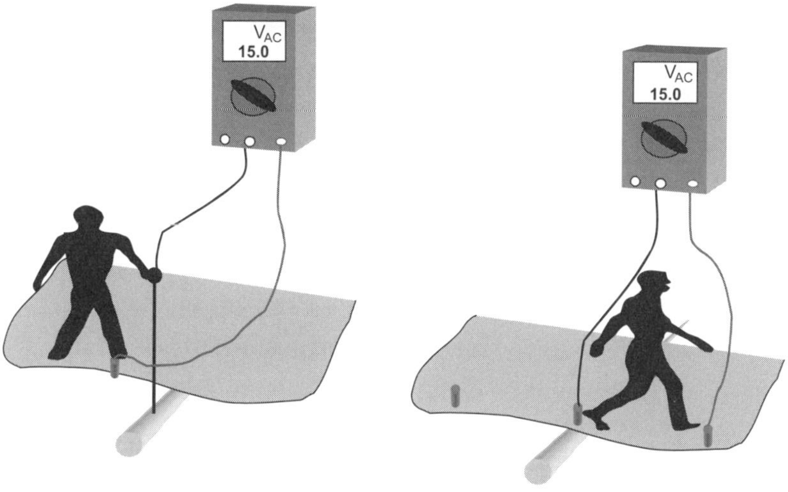

A. Touch Voltage **B. Step Voltage**

Figure 1.1 Touch and step voltages are shown.

current through the body and if the structure can be safely touched (touch voltage) or if the voltage from one foot to the other while walking (step voltage) is safe (Figure 1.1).

2.0 TOOLS AND EQUIPMENT

The following equipment will vary, depending on the test selected:

- Voltmeter capable of measuring 400 V_{AC}
- Copper-copper sulfate reference electrode or reference ground rod
- Test wires as necessary for the applicable test
- AC clamp-on ammeter
- Small hand tools with insulated handles

3.0 SAFETY EQUIPMENT

- Standard safety equipment and clothing, as required by the company's safety manual and regulations

- Electrically insulated clips and probe handles for meter leads
- Electrically approved gloves

4.0 PRECAUTIONS

The following precautions are in addition to those that must also be followed when working on a particular facility.[7.7,7.8]

4.1 Only personnel who have received training on the measurement of hazardous AC voltages and are qualified in accordance with local codes and regulations are to work on hazardous AC voltages or their supply.

4.2 When working near high-voltage AC (HVAC) power lines, take AC structure-to-ground voltage readings at frequent intervals because these voltages can change with the power line load and geometry.

4.3 Be aware that a wire extended parallel to an AC power line may have an induced AC voltage.

4.4 In areas of hazardous AC voltages, secure any exposed electrical terminals in a locked enclosure when the terminals are not attended, including test station wire terminations.

4.5 Measure a structure AC voltage to ground on the structure before taking cathodic protection (CP) measurements.

4.6 If the AC voltage to ground is equal to or exceeds 15 V_{AC}, practice the appropriate safety measures, advise other personnel working on the structure of the hazard, and inform management or the owner that mitigation is required.[7.7,7.8]

4.7 Be aware of the discontinuity locations between the pipeline and the power line because they will probably be the points of highest induced AC voltage.

4.8 Do not work on the structure when lightning is in the area.

4.9 When working near a fence or metal structures that parallel a HVAC power line, be aware that an AC voltage may be induced on the object.

4.10 Take precautions for local hazards such as traffic or dangerous animals and insects.

4.11 Only personnel qualified in AC stray current testing and analysis are to be responsible for these tests.

5.0 PROCEDURE

5.1 Voltmeters

5.1.1 Analog Voltmeters

5.1.1.1 The analog voltmeter is easily recognized by a readout consisting of a scale and a moving needle.[1] Hybrid meters that are electronic with an analog meter are also available.

5.1.1.2 A single-setting AC voltmeter will read at the indicated scale value. The full-scale value of a meter with multiple scale settings must be determined based on the scale setting. Read the scale setting and determine the actual full-scale value.

5.1.1.3 Determine the value of each scale major and minor division based on the number of divisions to the full-scale value.

5.1.1.4 Determine the number of divisions between zero (0) and the point where the needle is resting and multiply by the value of the division for the actual reading. Figure 5.1 is shown as a demonstration.

5.1.1.5 Using the example in Figure 5.1, the full-scale value is 40 V_{AC}, as indicated by the multiple-scale setting. (In this case, the dial is not a direct multiplier but indicates the full-scale value is 40 V so the scale reading is multiplied by 10.) Each major division is 10 V_{AC}, and each minor division is 2 $V_{AC.}$ The needle is resting 5.5 divisions above zero. The reading is then (5.5 divisions × 2 V_{DC}/division) = 11 V_{AC}, where AC stands for alternating current.

5.1.1.6 Alternately, use the major divisions to advantage. The major division below the needle is 10 V_{DC}, as determined in Section 5.1.1.5, and the needle is resting one-half division above it. The reading is then (10 V_{AC} + [0.5 division × 2 V_{DC}/division]) = 11 V_{AC}.

5.1.2 Chart Recorders

5.1.2.1 A chart recorder may be used to note the AC voltage variation over time at one location.

---

[1] *See* Cathodic Protection Procedure No. 2: Structure-to-Electrolyte Potential Measurement.

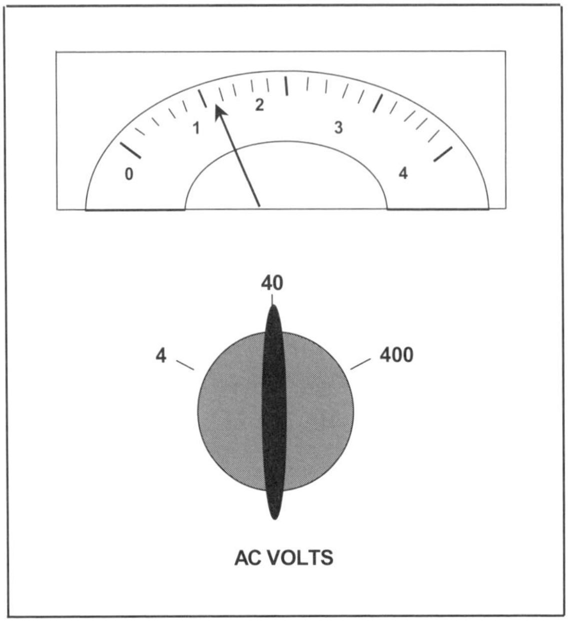

Figure 5.1 Sample analog AC voltmeter readout is shown.

5.1.2.2 The chart itself is the readout scale and is read similarly to an analog meter.

5.1.3 Digital Alternating Current Voltmeters

5.1.3.1 A digital AC voltmeter is an electronic instrument with a digital readout, normally an LCD. Note that the LCD is subject to freezing. Digital voltmeters may come with other features such as DC volts, an ohmmeter, diode testing, DC amperes [A], DC milliamperes, and/or DC microamperes. Make certain that it is set on the AC volts setting, which is sometimes represented by a sine wave symbol (**V ∿**)

5.1.3.2 The AC voltmeter scale may be manually adjusted, or it may have an autoranging scale, hold, and hertz features.

5.1.3.3 Always set the scale of a manually adjusted meter to the highest voltage setting and then reduce to the lowest scale that will capture the value of the reading. Note the units being measured on the display (volts or millivolts).

5.1.3.4 Digital voltmeters display the units and polarity, and those with an autoranging feature will automatically change the scale based on the value. When the measurement value exceeds a scale, the instrument will automatically change the scale and possibly the units, which may go unnoticed. For example, it may switch from a millivolt scale to a volt scale. It is therefore critical that the units and polarity be noted each time. (Polarity will not be indicated in AC.)

5.1.3.5 Use the hold feature to capture a reading so that it can be recorded afterward, especially in awkward locations. Be certain to release the hold for subsequent readings.

5.1.4 Data Logger

5.1.4.1 The data logger is a digital voltmeter that is normally programmed to enter AC voltage measurements at preset time intervals into a memory. These data, when downloaded into a computer, can be presented in a profile form.

5.1.4.2 The features within a data logger vary from one instrument to another and will be described in the manufacturer's operating manual. They are normally programmed to enter measurements on command or at time intervals. Although the potentials may be time stamped and appear to be accurate to the second, this must be confirmed. For example, when synchronizing data loggers, the display may be displaying the time to the same second, but one instrument may be at the start of the second, while the other may be near the end of the same second. In fact, the data of the one data logger will be recorded as 1 second (s) different.

5.2 Alternating Current Structure-to-Earth Voltage Measurements

5.2.1 Measure the AC pipe-to-earth voltage between the structure and the ground electrode, as shown in Figure 5.2. Although a standard CP reference electrode can be used, unlike a DC measurement, the metallic material of the reference electrode is

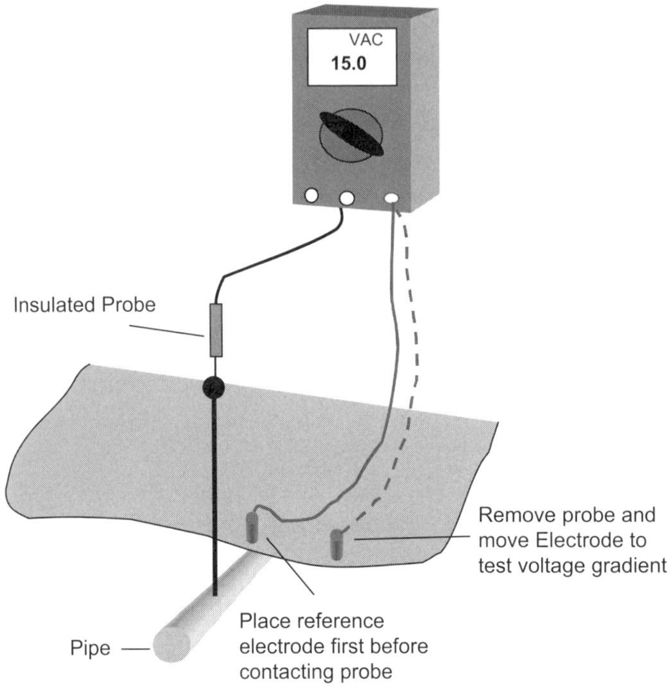

Figure 5.2 Typical AC pipe-to-earth voltage measurement is shown.

not important, and it need not be in a special electrolyte. Because AC voltage measurements are often made in conjunction with DC measurements for CP, a copper-copper sulfate electrode is often used for the AC pipe-to-earth voltage as it is required for the DC structure-to-electrolyte potential.

5.2.2 Connect a wire lead between the AC voltmeter common terminal and the reference electrode.

5.2.3 Connect a wire with an insulated probe to the second voltmeter terminal.

5.2.4 First, place the reference electrode into the ground.

5.2.5 Turn on the meter to the AC volts setting and, if not autoranging, set it to the highest scale.

5.2.6 Contact the structure or test lead wire by holding the insulated probe and record the AC volts by noting the value; if using a digital meter, note the units displayed (Figure 5.2).

5.2.7 Adjust the voltmeter scale to read on the lowest possible setting.

5.2.8 Remove the insulated probe, move the electrode 1 m (~3 ft) perpendicular to the pipeline, and retake the measurement. If the measurement is significantly different, repeat this measurement 2 m (~6 ft) perpendicular to the pipeline. (This is an area where a person may be standing while making contact with a pipeline appurtenance.)

5.2.9 Determine the percentage of total load under which the power line is operating during the AC pipe-to-ground voltage measurements. (The AC voltage at full power load should then be predicted.)

5.2.10 Report any values that are 15 V_{AC}[7.7,7.8] or greater and advise operating personnel that a hazardous AC voltage exists.

5.3 Testing of Alternating Current Mitigation Devices

5.3.1 Electrical Grounds

5.3.1.1 All measurements are to be made without human contact to any exposed terminals. Do not make contact with the terminals if disconnecting cables. Always connect the electrical ground before connecting to the structure.

5.3.1.2 Measure the AC voltage between the structure and the electrical ground, or across the DC decoupler (see Section 5.3.2), if one is installed (Figure 5.3).

5.3.1.3 Measure the AC current with a clamp-on ammeter (Figure 5.3).

5.3.1.4 Measure the AC earth gradient around the electrical ground by taking voltage measurements in a grid, as shown by the electrode locations in Figure 5.3.

5.3.1.5 Compare with previous data to ensure that the electrical ground resistance is similar and that it is performing as intended.

5.3.2 DC Decouplers

A DC decoupler is a device that allows AC to pass but blocks DC at CP DC voltage levels. Two types are available, including an electrolytic type and an electronic type (Figures 5.4 and 5.5).

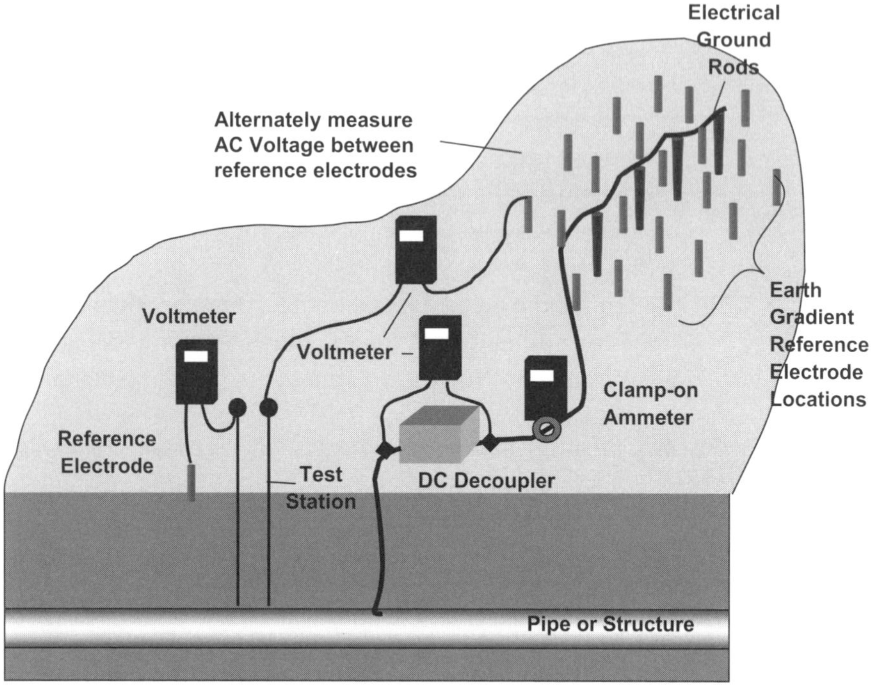

Figure 5.3 Measuring AC earth gradient over electrical ground is shown.

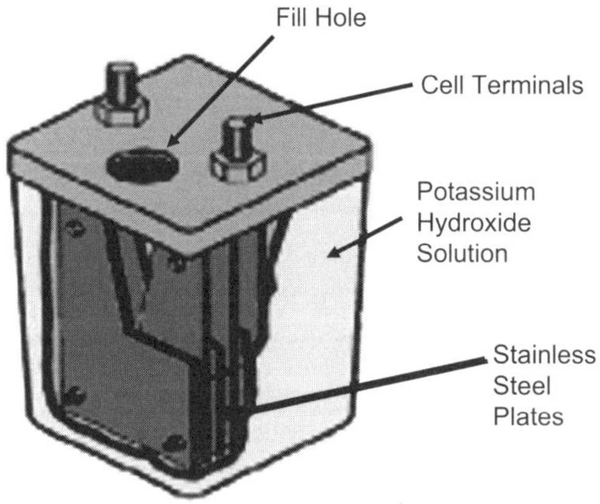

Figure 5.4 Electrolytic DC decoupler is shown. From NACE CP Technician course.

Figure 5.5 Electronic DC decoupler is shown. From NACE CP Technician course.

5.3.2.1 Electrolytic Decouplers

5.3.2.1.1 Measure the AC voltage across the electrolytic DC decoupler, that is, between the structure and the electrical ground.

5.3.2.1.2 Measure the AC current through the electrolytic DC decoupler with an AC clamp-on ammeter.

5.3.2.1.3 Inspect the cables and connections to ensure that they are in good condition.

5.3.2.1.4 Measure the DC voltage across the electrolytic DC decoupler, that is, between the structure and the electrical ground.

5.3.2.1.5 Measure the DC current through the electrolytic DC decoupler with a DC clamp-on ammeter.

5.3.2.1.6 Inspect the plates in the cell to confirm that they are in good condition.

5.3.2.1.7 Inspect the potassium hydroxide solution to make certain that it is at the proper fill level.

5.3.2.1.8 Confirm that there is a film of oil on the surface of the potassium hydroxide.

5.3.2.1.9 Confirm that the measurements are within the rating of the electrolytic DC decoupler.

5.3.2.2 Electronic Decouplers

5.3.2.2.1 Measure the AC voltage across the electronic DC decoupler, that is, between the structure and the electrical ground.

5.3.2.2.2 Measure the AC current through the cables to the electronic DC decoupler with an AC clamp-on ammeter.

5.3.2.2.3 Inspect the cables and connections to ensure that they are in good condition.

5.3.2.2.4 Measure the DC voltage across the electronic DC decoupler, that is, between the structure and the electrical ground.

5.3.2.2.5 Measure the DC current through the electronic DC decoupler with a DC clamp-on ammeter.

5.3.2.2.6 Inspect the components in the DC decoupler for signs of arcing or damage.

5.3.2.2.7 Confirm that the measurements are within the rating of the electronic DC decoupler.

5.3.3 Alternating Current Voltage Gradient Grids

5.3.3.1 An AC voltage gradient grid is intended to bring a person to the voltage of the structure as he or she approaches it so that the voltage difference between the structure and the ground is safe when contact is made. Because the grid is installed close to the surface, it is not intended to reduce the AC voltage, as is the case with an electrical ground.

5.3.3.2 Connect a wire lead between the AC voltmeter common terminal and the reference electrode.

5.3.3.3 Connect a wire with an insulated probe to the second voltmeter terminal.

5.3.3.4 First, place the reference electrode into the ground near the center of the voltage gradient grid.

5.3.3.5 Contact the structure or test lead wire by holding the insulated probe and record the AC volts. Remove the insulated probe, move the electrode 1 m (~3 ft) perpendicular to the pipeline, and retake the measurement. Remove the insulated probe and repeat this

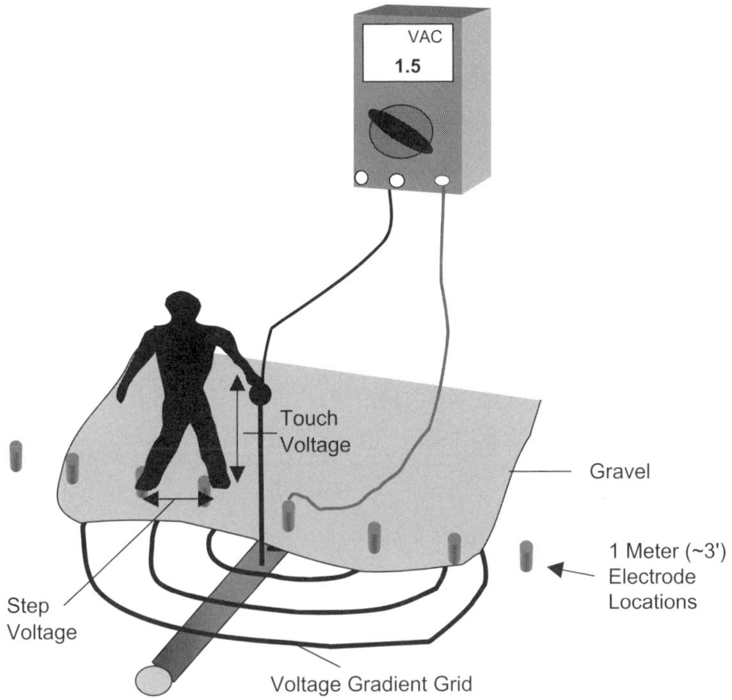

Figure 5.6 Testing a voltage gradient grid is shown.

measurement with the reference electrode 2 m (~6 ft) perpendicular to the pipeline. Continue these readings at 1-m (~3-ft) intervals (Figure 5.6) to a point 1 m beyond the edge of the gravel extending beyond the grid. Repeat across another part of the grid.

5.3.3.6 If a sacrificial anode material is used for the gradient grid at a CP test station, the grid would have to be disconnected to measure a true polarized structure-to-electrolyte potential. This, in effect, disconnects a safety feature; therefore, another method has to be used to determine polarized structure-to-electrolyte potentials. The cable connecting the grid must be disconnected using electrically insulated tools and/or electrically approved gloves until an AC structure-to-earth voltage can be measured to confirm that it is safe.

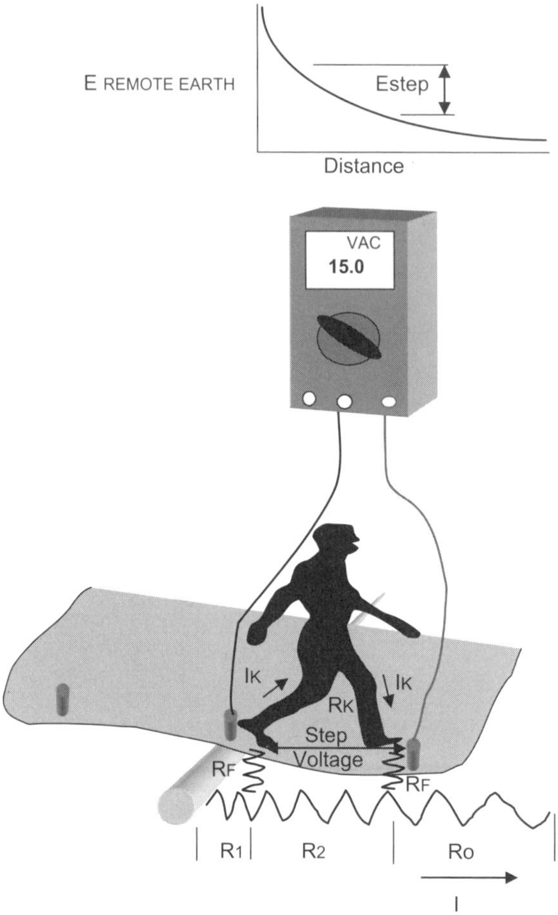

Figure 6.1 Step voltage is shown.

6.0 ANALYSIS

6.1 Step and Touch Voltages

6.1.1 Step Voltage

A step voltage encountered by a person walking across a voltage gradient is illustrated by Figure 6.1, and the equivalent electrical circuit is shown in Figure 6.2.

Dalziel concluded that 99.5% of all persons could withstand a current through their body as defined by Equation (6.1),

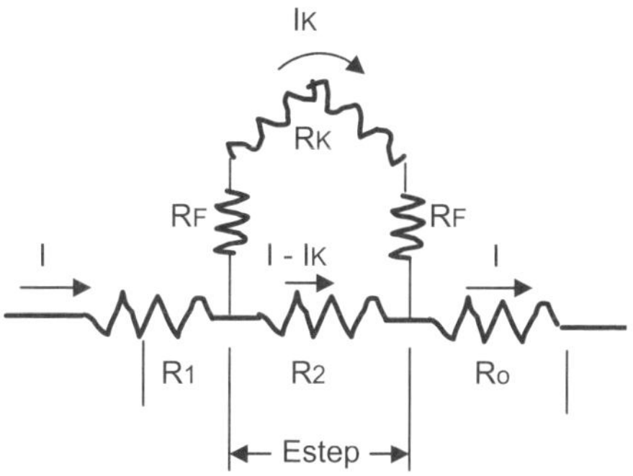

Figure 6.2 Equivalent electrical step voltage circuit is shown.

without fibrillation[7,9]:

$$I_K = \frac{0.116}{\sqrt{t}}, \tag{6.1}$$

where

- I_K rms current through body (amperes; see Figures 6.1 and 6.2)
- t time of exposure (seconds)

I_K from Ohm's law is given by Equation (6.2):

$$I_K = \frac{E_{\text{step}}}{(R_K + 2R_F)} \tag{6.2}$$

where

- E_{step} tolerable potential difference between any two points on the ground that can be contacted by the feet
- R_K resistance of the body
- R_F resistance of a foot

Combining Equations (6.1) and (6.2) gives Equation (6.3):

$$E_{\text{step}} = \frac{0.116}{\sqrt{t}}(R_K + 2R_F). \tag{6.3}$$

The resistance of a person's foot to soil is given by Equation (6.4)[7.7]:

$$R_F = 3\rho, \tag{6.4}$$

where

ρ soil resistivity (ohm-meters)

Therefore, the step voltage can be calculated by Equation (6.5):

$$E_{\text{step}} = \frac{0.116}{\sqrt{t}}(R_K + 6\rho). \tag{6.5}$$

A conservative value for a body resistance is 1,000 Ω. With $R_K = 1{,}000\ \Omega$, Equation (6.5) becomes Equation (6.6), or is simplified to Equation (6.7):

$$E_{\text{step}} = \frac{0.116}{\sqrt{t}}(1{,}000 + 6\rho) \tag{6.6}$$

$$E_{\text{step}} = \frac{(116 + 0.7\rho)}{\sqrt{t}}. \tag{6.7}$$

The preceding equations and a further discussion are also covered in IEEE Standard 80.[7.10]

6.1.2 Touch Voltage

A touch voltage that a person may encounter between the point of body contact and the earth is illustrated by Figures 6.3 and 6.4.

The equation for I_K during touch voltage is given in Equation (6.8):

$$I_K = \frac{E_{\text{touch}}}{\left(R_K + \frac{R_F}{2}\right)}, \tag{6.8}$$

where

I_K rms current through body (amperes; see Figures 6.1 and 6.2)

E_{touch} tolerable potential difference between any point of the ground where a person may stand and any point that can be touched simultaneously (volts)

R_K resistance of the body (ohms)

R_F resistance of a foot (ohms)

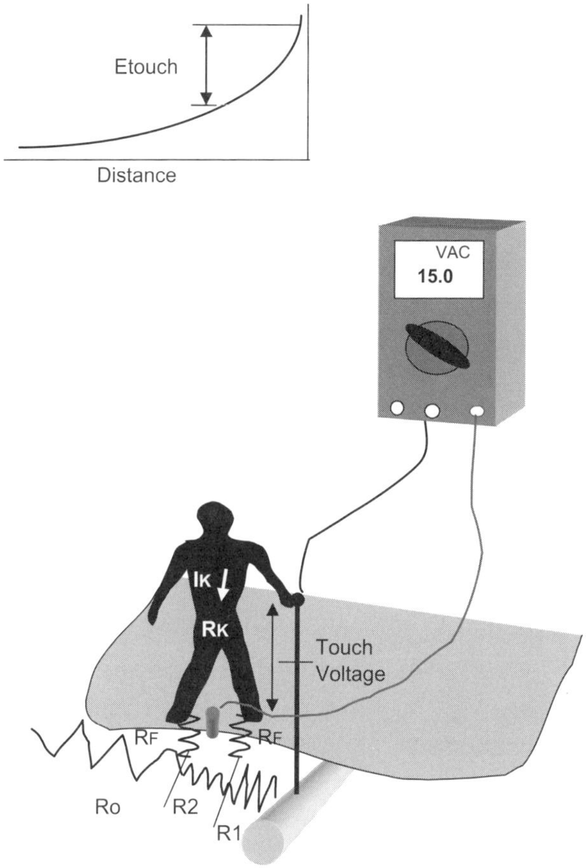

Figure 6.3 Touch voltage is shown.

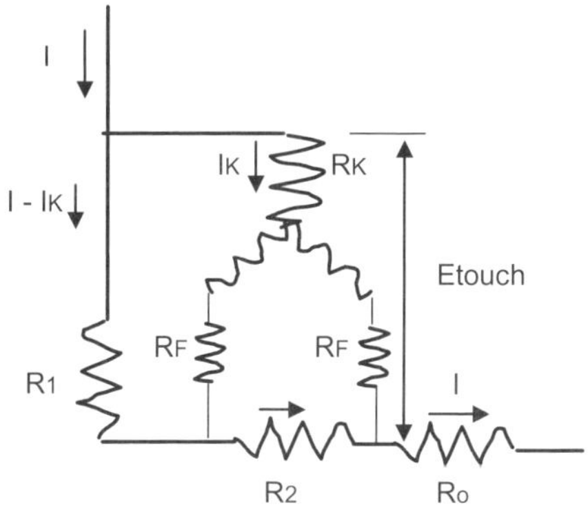

Figure 6.4 Equivalent electrical circuit for touch voltage near a grounded structure is shown.

Combining Equations (6.1) and (6.8) gives Equation (6.9):

$$E_{\text{touch}} = \frac{0.116}{\sqrt{t}}\left(R_K + \frac{R_F}{2}\right). \tag{6.9}$$

With $R_F = 3\rho$, Equation (6.9) becomes Equation (6.10):

$$E_{\text{touch}} = \frac{0.116}{\sqrt{t}}(R_K + 1.5\rho). \tag{6.10}$$

Using a conservative body resistance (R_K) of 1,000 Ω, Equation (6.10) becomes Equations (6.11) and (6.12):

$$E_{\text{touch}} = \frac{0.116}{\sqrt{t}}(1{,}000 + 1.5\rho) \tag{6.11}$$

$$E_{\text{touch}} = \frac{(116 + 0.17\rho)}{\sqrt{t}}. \tag{6.12}$$

These equations and a further discussion are included in IEEE Standard 80.[7.10]

6.2 Alternating Current Structure-to-Earth Voltage Measurements

6.2.1 Note that the preceding equations for safe step and touch voltages include a short time exposure factor that applies to the duration of a fault.

6.2.2 Under steady-state conditions, an AC voltage of 15 V_{AC} or greater is considered a hazardous voltage, such as with induced AC voltages.[7.7,7.8] This contact may be directly to the structure or through a metal object such as a rod, small pipe, or conduit; therefore, the AC voltage measured at a point remote from the structure must also be considered. In areas where young children may come in contact with a structure, a lower value than 15 V_{AC} must be used as a safe AC voltage.[7.7,7.8]

6.2.3 A source current capacity of 5 mA or more is also considered hazardous.[7.7] This gives a shock sensation but is still within let-go muscular control. (Loss of muscular control and painful shock happens at 15 mA.) The current can be estimated from an anticipated human resistance under different conditions and the

Table 6.1 Human Resistance to Electrical Current[7.7]

Conditions	Resistance (Ω)
Dry skin	100,000–600,000
Wet skin	1,000
Internal body: hand to foot	400–600
Internal body: ear to ear	~100

AC voltage. An estimation of human resistance is shown in Table 6.1.

Table 6.1 illustrates that if body contact occurs under wet conditions, a lower value than 15 V_{AC} should be adopted as a hazardous level.

6.2.4 Higher AC voltages are more likely at discontinuities between the pipeline and the power line than in between. Make certain that the AC voltage measurements were taken at the discontinuity and not at a convenient test station nearby. Otherwise, the results may not show the greatest exposure.

7.0 REFERENCES

7.1 M.A. Puschell, "Power Lines and Pipelines in Close Proximity During Construction and Operation," *MP* 12, 12 (1973): pp. 28–32.

7.2 E.A. Cherney, "Pipeline Voltage Hazards on High Voltage AC Transmission Line Right-of-Way," *MP* 14, 3 (1975): pp. 29–33.

7.3 A.W. Hamlin, "Some Effects of Alternating Current on Pipeline Operation," *MP* 19, 1 (1980).

7.4 BC Hydro and Power Authority, *Study of Problems Associated with Pipelines Occupying Joint-Use Corridors with AC Transmission Lines* (Ottawa, ON: Canadian Electric Association, 2000).

7.5 J. Dabkowski, A. Taflove, *Mutual Design Considerations for Overhead AC Transmission Lines and Gas Transmission Pipelines* (Washington, DC: American Gas Association, and Palo Alto, CA: Electric Power Research Institute, 1978).

7.6 M. Frazier, *Power Line–Induced AC Potential on Natural Gas Pipelines for Complex Right-of-Way Configurations* (Washington, DC: American Gas Association, and Palo Alto, CA: Electric Power Research Institute, 1983).

7.7 NACE Standard SP0177-2007, "Mitigation of Alternating Current and Lightning Effects on Metallic Structures and Corrosion Control Systems" (Houston, TX: NACE International, 2007).

7.8 CSA 22.3 No. 6, "Principles and Practices of Electrical Coordination Between Pipelines and Electric Supply Lines" (Rexdale, Toronto, ON: Canadian Standards Association).

7.9 C.F. Dalziel, "Electric Shock Hazard," *IEEE Spectrum* 9 (1972): p. 41–50.

7.10 IEEE Standard 80, "IEEE Guide for Safety in Substation Grounding" (New York: IEEE Standards Board).

CATHODIC PROTECTION PROCEDURE NO. 12

Soil Resistivity Measurement 12

1.0 INTRODUCTION

The purpose of this procedure is to determine the resistivity of an electrolyte either by a Wenner four-pin method or by a soil box.[7.1] A modification to the four-pin method is also described, to be used where space is limited.[7.2]

Soil resistivity measurements are important both in the prediction of a corrosive electrolyte and in the design of cathodic protection systems.[7.3–7.5]

The four-pin method is used in soils, while the soil box can be used for samples of soils and liquids. Note that the four-pin test measures the soil resistivity to an average depth that is the same as the spacing between the two inside pins. A factor must be determined to convert a resistance measurement in ohms to resistivity in ohm-centimeters or ohm-meters.

2.0 TOOLS AND EQUIPMENT

2.1 Option 1: Direct Current Four-Pin Measurement—Equal Pin Spacing

- Direct current (DC) ammeter capable of measuring from 10 mA to 10 A, complete with leads
- High-impedance DC voltmeter capable of measuring 1–4,000 mV and 20 V, complete with leads
- Two test wires, complete with a clip on one end, that are at least 1.5 times the maximum spacing between the pins
- Two test wires, complete with a clip on one end, that are greater than 0.5 times the maximum spacing between the inside pins

- Four pins that are pointed on one end; a handle on each pin is optional but convenient
- DC power supply with a variable voltage output or battery and a control resistor
- Double-pole, double-throw (three positions, with center off) switch to interrupt or reverse polarity

2.2 Option 2: Alternating Current Four-Pin Measurement

- Soil resistivity instrument (alternating current [AC]), complete with leads
- Two test wires, complete with a clip on one end, that are at least 1.5 times the maximum spacing between the pins
- Two test wires, complete with a clip on one end, that are greater than 0.5 times the maximum spacing between the inside pins
- Four pins that are pointed on one end; a handle on each pin is optional but convenient
- Note that if several measurements are to be taken at the same pin spacing, a harness can be assembled with clips at the desired spacing

2.3 Option 3: Direct Current Four-Pin Measurement—Unequal Pin Spacing

- DC ammeter capable of measuring 10 mA to 10 A, complete with leads
- High-impedance DC voltmeter capable of measuring 1–4,000 mV and 20 V, complete with leads
- Two test wires complete with a clip on one end that are at least 1.5 times the maximum spacing intended between the pins
- Two test wires complete with a clip on one end that is greater than 0.5 the maximum spacing intended between the inside pins
- Four pins that are pointed on one end. A handle on each pin is optional but convenient
- DC power supply with a variable voltage output or battery and a control resistor
- Double-pole, double-throw (three positions, with center off) switch to interrupt or reverse polarity

2.4 Option 4: Direct Current Soil (Water) Box

- DC ammeter capable of measuring 1.0 mA to 1.0 A, complete with leads
- High-impedance DC voltmeter capable of measuring 1–4,000 mV and 20 V, complete with leads
- Soil resistivity box of known calibration
- DC power supply with a variable voltage output
- Double-pole, double-throw (three positions, with center off) switch to interrupt or reverse polarity

2.5 Option 5: Alternating Current Four-Pin Soil (Water) Box

- Soil resistance (resistivity) instruments, complete with test wires
- Four-pin soil resistivity box of known calibration

2.6 Option 6: Alternating Current Two-Pin Soil (Water) Box

- Soil resistivity instruments, complete with test wires
- Two-pin soil resistivity box of known calibration

2.7 Option 7: Resistivity Probe

- Soil resistivity instrument, complete with test wires and probe

3.0 SAFETY EQUIPMENT

- Standard safety equipment, as required by the company's safety manual and regulations
- Electrically insulated clips and insulated pin handles if near high-voltage AC power lines
- Be aware that a high voltage exists between the C1 and C2 terminals of some AC resistivity meters when they are operating

4.0 PRECAUTIONS

4.1 Ensure that there are low contact resistances between the instruments and wires and between the wires and the pins.

4.2 Pins are to be installed with the tips in moist soil.

4.3 Do not touch any exposed current (C1 and C2) connections or pins while the test is in progress or when the soil resistivity instrument is in operation because high voltage can be present.

4.4 Do not run the pins in close proximity to a parallel metallic structure or on top of large-diameter pipes or tanks because their influence will invalidate the results.

4.5 Where soil resistivity to deeper depths is required, the conventional soil resistivity instruments may not be sensitive enough to measure low resistance values, and thus the resistivity value cannot be calculated. Option 1, DC four-pin measurement, may then be necessary, instead of option 2, AC four-pin measurement.

4.6 Where there is not enough clear surface area or where the test sensitivity is poor, option 3 can be considered.

5.0 PROCEDURE

Section 5.0 covers testing procedures, while Section 6.0 provides the calculations for resistivity based on the resistance values measured in this section.

5.1 Option 1: Direct Current Four-Pin Measurement

5.1.1 Determine the depth at which the soil resistivity is to be averaged.

5.1.2 Insert four pins into the ground at equal spacing that is equal to the depth to be measured, as determined previously.

5.1.3 Label the pins from one end to the other as C1, P1, P2, and C2, as shown in Figure 5.1.

5.1.4 Connect one of the longer wires to the C1 pin and the other end to one of the terminals of the two-pole, double-throw switch (Figure 5.1).

5.1.5 Connect the other, longer wire to the C2 pin and the other end of the wire to one terminal of the ammeter, and the other terminal of the ammeter is to be connected to the other terminal of the two-pole, double-throw switch (Figure 5.1).

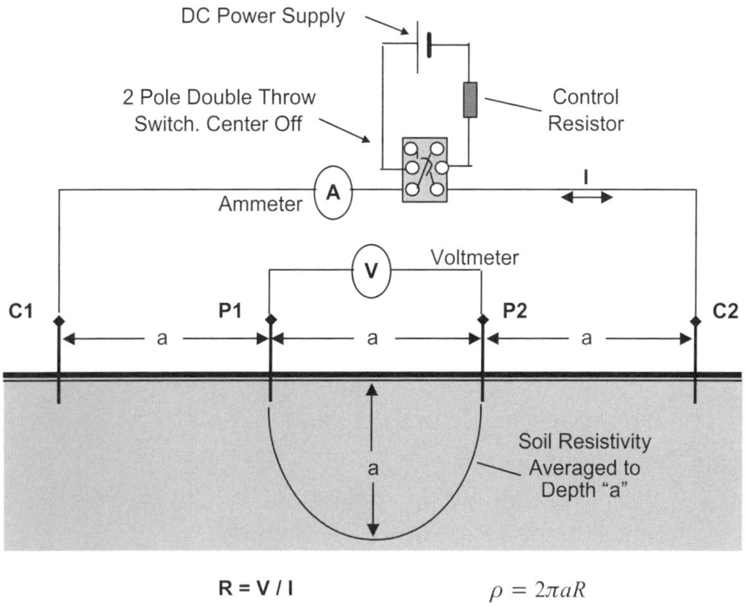

Figure 5.1 DC four-pin soil resistivity test with equal spacing is shown.

5.1.6 Connect the power supply terminals to the center terminals of the switch such that the positive side of the DC power supply will be connected through the switch to the C1 pin (Figure 5.1). Do not energize.

5.1.7 Connect the positive of the voltmeter to the P1 pin and the negative to the P2 pin.

5.1.8 Measure the potential (*Vo*) between the P1 and P2 pins without current and allow to stabilize.

5.1.9 Be prepared to take a potential and current measurement within a short time after energizing the DC power supply and then to turn it off. The pins will polarize if the current is left on, and the measurements will change.

5.1.10 Energize the DC power supply, and within a few seconds, quickly measure and record the potential ($V_{1\text{ on}}$) between the P1 and P2 pins and the current (I_1) applied. Note the polarity of each reading.

5.1.11 Immediately interrupt the current and measure the OFF potential ($V_{1\text{ off}}$) between the P1 and P2 pins.

5.1.12 Reverse the polarity and repeat the steps given in Sections 5.1.9, 5.1.10, and 5.1.11. Identify these measurements as $V_{2\ \text{on}}$, $V_{2\ \text{off}}$, and I_2.

5.1.13 Repeat the test detailed in Sections 5.1.9 to 5.1.12 at a different current value.

5.2 Option 2: Alternating Current Four-Pin Measurement

5.2.1 Determine the depth at which the soil resistivity is to be averaged.

5.2.2 Insert four pins into the ground at equal spacing that is equal to the depth to be measured, as determined previously.

5.2.3 Label the pins from one end to the other as C1, P1, P2, and C2, as shown in Figure 5.2.

5.2.4 Connect one of the longer wires from the C1 pin to the C1 terminal of the soil resistivity instrument.

5.2.5 Connect the other, longer wire from the C2 pin to the C2 terminal of the soil resistivity instrument.

5.2.6 Connect one of the shorter wires from the P1 pin to the P1 terminal of the soil resistivity instrument, which is on the same side as C1.

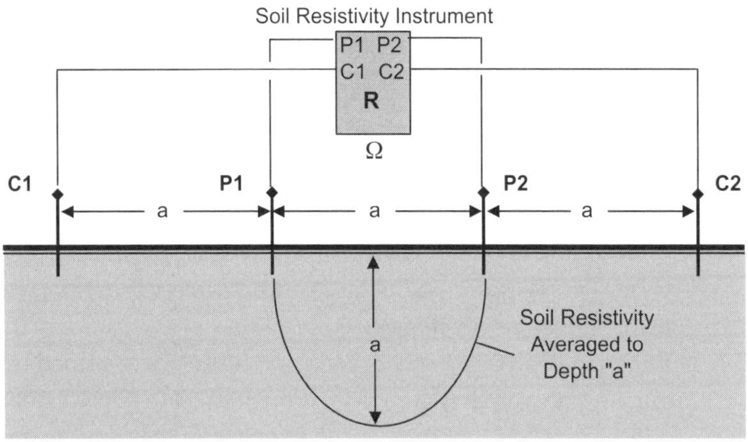

$$\rho = 2\pi aR$$

Figure 5.2 AC four-pin soil resistivity test is shown.

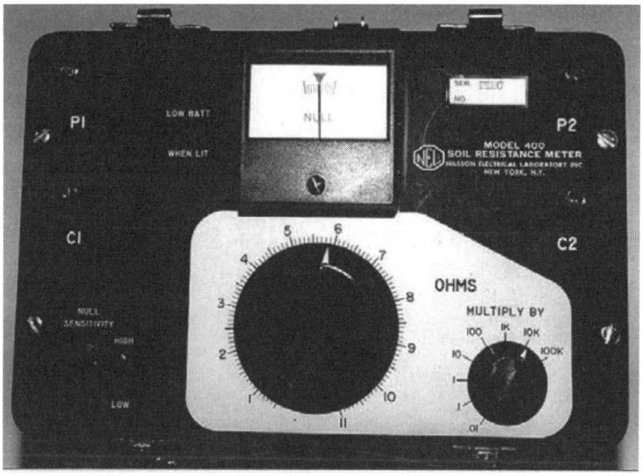

Figure 5.3 AC soil resistivity instrument panel is shown. Photo from the NACE CP Technician course.

5.2.7 Connect the other, shorter wire from the P2 pin to the P2 terminal of the soil resistivity instrument, which is on the same side as C2. The instrument will not balance if the potential (P1 and P2) wires are reversed.

5.2.8 Balance the instrument as shown in Figure 5.3. If unfamiliar with the instrument, the following procedure is suggested.

5.2.8.1 Rotate the balancing dial and the multiplier dial clockwise to the highest setting.

5.2.8.2 While temporarily holding the switch to the low-sensitivity setting, the meter needle will swing to the right. Rotate the multiplier dial counterclockwise until the needle swings to the left, and then rotate the multiplier dial one setting clockwise. Release the switch.

5.2.8.3 While touching the switch to the low-sensitivity position, rotate the dial until the needle does not move right or left from the null position when the switch is on.

5.2.8.4 Repeat the procedure in Section 5.2.8.3 while touching the switch to the high-sensitivity position.

5.2.9 Record the value to which the dial is pointing on the graduated scale and the multiplier value. The resistance is the product of these values:

$$R = \text{Dial} \times \text{Multiplier},$$

where

R	resistance (ohms)
Dial	center adjustment dial in Figure 5.3
Multiplier	lower right-hand multiplier dial in Figure 5.3

5.3 Option 3: Four-Pin Unequal Spacing

5.3.1 Although equal spacing is preferred, this method may be used where a loss of sensitivity has occurred during wide pin spacing measurements and is a modification of options 1 and 2. Note that a soil resistivity meter may be substituted for the power supply, voltmeter, and ammeter as between options 1 and 2 if this pin spacing falls within its sensitivity.

5.3.2 Determine the depth at which the soil resistivity is to be averaged and insert the two inside potential pins (P1 and P2) at a spacing equal to that depth.

5.3.3 Insert the two current pins (C1 and C2) into the ground at the same spacing from their potential pins, although at a shorter spacing than between the potential pins.

5.3.4 Label the pins from one end to the other as C1, P1, P2, and C2, as shown in Figure 5.4.

5.3.5 Connect one of the longer wires to the C1 pin and the other end to one of the terminals of the two-pole, double-throw switch (Figure 5.4).

5.3.6 Connect the other, longer wire to the C2 pin and the other end of the wire to one terminal of the ammeter, and the other terminal of the ammeter is to be connected to the other terminal of the two-pole, double-throw switch (Figure 5.4).

5.3.7 Connect the power supply terminals to the center terminals of the switch such that the positive side of the DC power supply will be connected through the switch to the C1 pin (Figure 5.4). Do not energize.

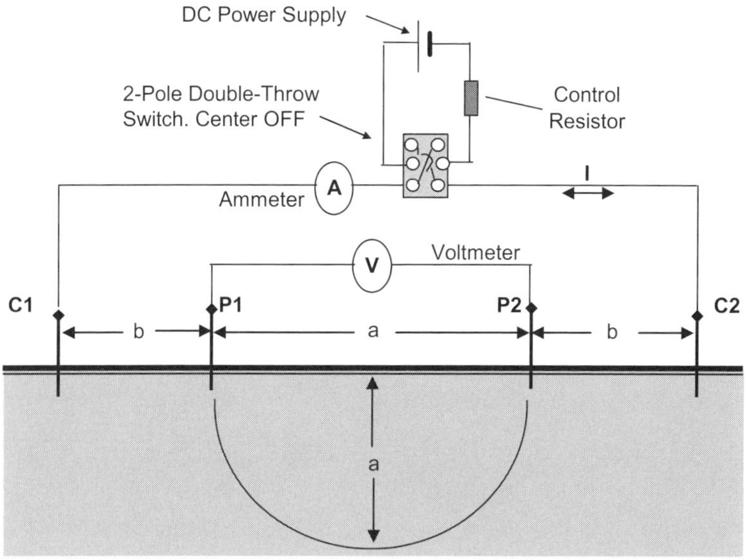

Figure 5.4 DC four-pin soil resistivity test with unequal spacing is shown.

5.3.8 Connect the positive of the voltmeter to the P1 pin and the negative to the P2 pin.

5.3.9 Measure the potential (*Vo*) between the P1 and P2 pins without current and allow to stabilize.

5.3.10 Be prepared to take a potential and current measurement within a short time after energizing the DC power supply and then to turn it off. The pins will polarize if the current is left on, and the measurements will change.

5.3.11 Energize the DC power supply, and within a few seconds, quickly measure and record the potential ($V_{1\ \mathrm{on}}$) between the P1 and P2 pins and the current (I_1) applied. Note the polarity of each reading.

5.3.12 Immediately interrupt the current and measure the OFF potential ($V_{1\ \mathrm{off}}$) between the P1 and P2 pins.

5.3.13 Reverse the polarity and repeat the steps given in Sections 5.3.10, 5.1.11, and 5.1.12. Identify these measurements as $V_{2\ \mathrm{on}}$, $V_{2\ \mathrm{off}}$, and I_2.

5.3.14 Repeat the test given in Sections 5.3.10–5.3.13 at a different current value.

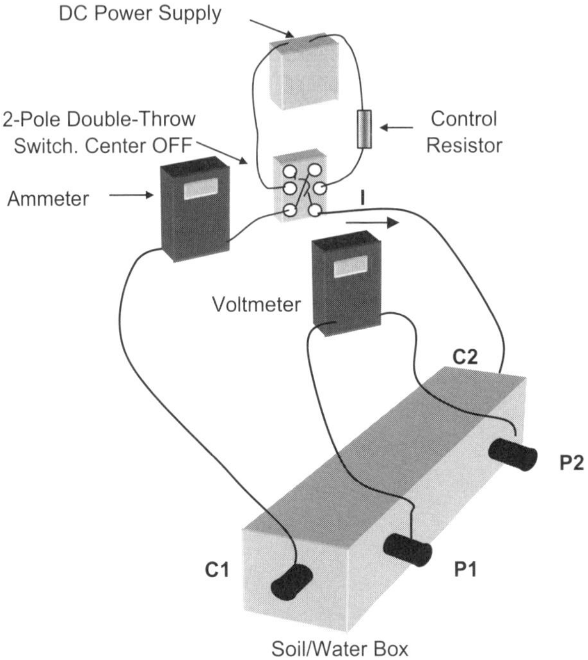

Figure 5.5 Four-terminal soil box DC test is shown.

5.3.15 If using a soil resistivity meter, replace the steps given in Sections 5.3.9–5.3.14 with those of Sections 5.2.8 and 5.2.9.

5.4 Option 4: Direct Current Soil Box (Four Terminals)

5.4.1 Select a representative soil (water) sample that is at least twice the volume of the soil box and seal in a plastic bag.

5.4.2 Completely fill the soil box with an appropriate representative sample of the soil (water). Pack the soil and remove all voids. Large rocks should be eliminated but noted in the records.

5.4.3 The four terminals relate to the four-pin method described in Section 5.1. The two end terminals are equivalent to C1 and C2, while the two inside terminals correspond to P1 on the left and P2 on the right (see Figure 5.5).

5.4.4 Connect one wire to the C1 terminal and its other end to one of the terminals of the two-pole, double-throw switch (Figure 5.5).

5.4.5 Connect another wire to the C2 terminal and to one terminal of the ammeter. The other terminal of the ammeter is to be connected to the other terminal of the two-pole, double-throw switch (Figure 5.5).

5.4.6 Connect the DC power supply terminals to the center terminals of the switch such that the positive side of the DC power supply will be connected through the switch to the C1 pin (Figure 5.5). Do not energize.

5.4.7 Connect the positive of the voltmeter to the P1 pin and the negative to the P2 pin.

5.4.8 Measure the potential (Vo) between the P1 and P2 pins without current and allow to stabilize.

5.4.9 Be prepared to take a potential and current measurement within a short time after energizing the DC power supply and then to turn it off. The plates will polarize if the current is left on, and the measurements will change.

5.4.10 Energize the DC power supply, and within a few seconds, quickly measure and record the potential ($V_{1\ \mathrm{on}}$) between the P1 and P2 pins and the current (I_1) applied. Note the polarity of each reading.

5.4.11 Immediately interrupt the current and measure the OFF potential ($V_{1\ \mathrm{off}}$) between the P1 and P2 pins.

5.4.12 Reverse the polarity and repeat the steps given in Sections 5.4.9, 5.4.10, and 5.4.11. Identify these measurements as $V_{2\ \mathrm{on}}$, $V_{2\ \mathrm{off}}$, and I_2.

5.4.13 Repeat the steps from Sections 5.4.9 to 5.4.12 at a different current value.

5.4.14 Calculate the resistance of the test in Sections 5.4.10, 5.4.11, and 5.4.12 as follows:

$$R_1 = \frac{(V_{1\ \mathrm{on}} - V_{1\ \mathrm{off}})}{I_1}$$

$$R_2 = \frac{(V_{2\ \mathrm{on}} - V_{2\ \mathrm{off}})}{I_2},$$

where

R_1, R_2	resistance of test 1 and test 2, respectively (ohms)
$V_{1\ on}$, $V_{2\ on}$	potential between P1 and P2, respectively, with current applied (volts)
$V_{1\ off}$, $V_{2\ off}$	potential between P1 and P2, respectively, with current off (volts)
I_1, I_2	current C1 and C2 (amperes)

5.4.15 Calculate the values for the tests at different current values in the same manner as detailed in Section 5.4.14.

5.4.16 If the resistance values are not similar, repeat the test until similar results are obtained.

5.4.17 Average the forward and reverse resistance values:

$$R = \frac{(R_1 + R_2)}{2},$$

where

R	resistance (ohms)
R_1, R_2	calculated resistance from test 1 and test 2, respectively (ohms)

See Section 6.0 for the calculation of soil resistivity.

5.5 Option 5: Alternating Current Soil Box (Four Terminals)

5.5.1 Select a representative soil (water) sample that is at least twice the volume of the soil box and seal in a plastic bag.

5.5.2 Completely fill the soil box with an appropriate representative sample of the soil (water). Pack the soil and remove all voids. Large rocks should be eliminated but noted in the records.

5.5.3 The four terminals relate to the four-pin method described in Sections 5.2 and 5.3. The two end terminals are equivalent to C1 and C2, while the two inside terminals correspond to P1 on the left and P2 on the right (see Figure 5.6).

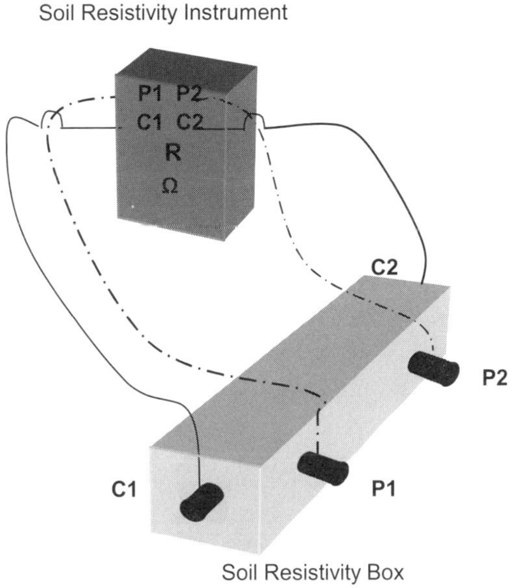

Figure 5.6 Four-terminal soil box AC test is shown.

5.5.4 Connect a wire from the C1 terminal on the soil box to the C1 terminal of the soil resistivity instrument.

5.5.5 Connect another wire to the C2 terminal on the soil box and to the C1 terminal of the soil resistivity instrument.

5.5.6 Connect a wire from the P1 soil box terminal to the P1 terminal of the soil resistivity instrument, which is on the same side as the C1 connection.

5.5.7 Connect another wire from the P2 soil box terminal to the P2 terminal of the soil resistivity instrument, which is on the same side as the C2 connection. The instrument will not balance if the potential (P1 and P2) wires are reversed.

5.5.8 Balance the instrument. If unfamiliar with the instrument, the following procedure is suggested.

5.5.8.1 Rotate the balancing dial and the multiplier dial clockwise to the highest setting.

5.5.8.2 While temporarily holding the switch to the low-sensitivity setting, the meter needle will swing to the right. Rotate the multiplier dial counterclockwise until the needle swings to the left, and then rotate the multiplier dial one setting clockwise. Release the switch.

5.5.8.3 While touching the switch to the low-sensitivity position, rotate the dial until the needle does not move right or left from the null position when the switch is on.

5.5.8.4 Repeat the procedure in Section 5.4.8.3 while touching the switch to the high-sensitivity position.

5.5.9 Record the value to which the dial is pointing on the graduated scale and the multiplier value. The resistance is the product of these values:

$$R = \text{Dial} \times \text{Multiplier},$$

where

R	resistance (ohms)
Dial	center adjustment dial (Figure 5.3)
Multiplier	lower right-hand multiplier dial (Figure 5.3)

5.6 Option 6: Alternating Current Soil Box (Two Terminals)

5.6.1 Select a representative soil (water) sample that is at least twice the volume of the soil box and keep in a sealed plastic bag.

5.6.2 Completely fill the soil box with an appropriate representative sample of the soil (water). Pack the soil and remove all voids. Large rocks should be eliminated but noted in the records.

5.6.3 The end of the two-terminal soil resistivity box is a heavy metal plate. Wires from each of the C1 and P1 soil resistivity instrument terminals are connected to one plate, while the C2 and P2 terminals are connected to the plate on the other end (Figure 5.7). Run a separate wire from each instrument terminal, rather than shorting the terminals at the instrument. If the P1 is bonded to the C1 at the instrument and, similarly, the P2 to the C2 terminals, then the wires to the box are included in the resistance measurements.

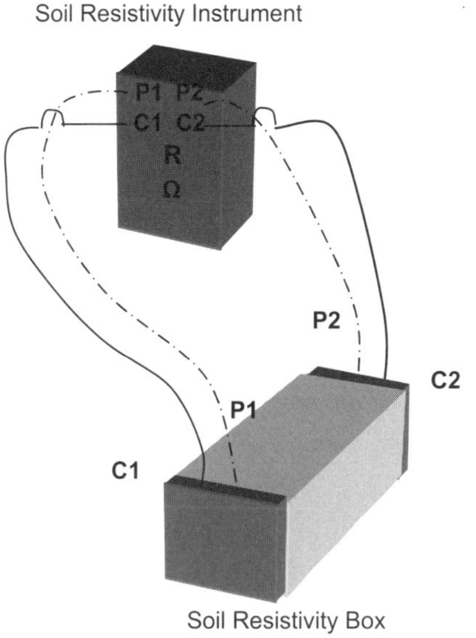

Figure 5.7 Two-terminal soil box AC test is shown.

5.6.4 Follow the instructions in Section 5.5.8.

5.6.5 Record the value at which the dial is pointing on the graduated scale and the multiplier value. The resistance is the product of these values:

$$R = \text{Dial} \times \text{Multiplier},$$

where

R	resistance (ohms)
Dial	center dial in Figure 5.3
Multiplier	lower right-hand dial in Figure 5.3

5.7 Option 7: Resistivity Probe

5.7.1 Adjust the instrument by the rotating knob until a null (reduction in sound beyond which the sound increases) is heard in the ear piece (Figure 5.8).

5.7.2 Before connecting the probe, confirm that the instrument is calibrated by turning the instrument on and balancing the

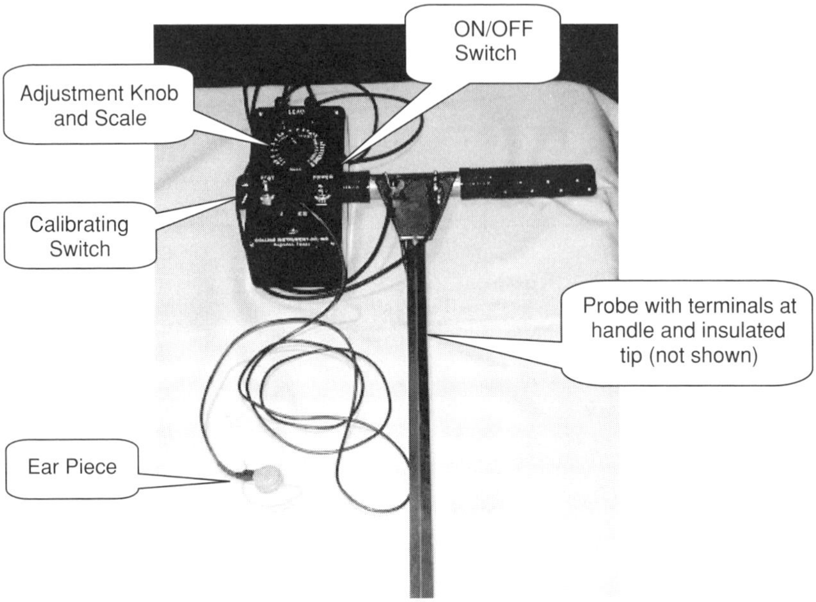

Figure 5.8 Resistivity probe and instrument is shown. Photo from NACE CP Tester course.

instrument while holding the calibrating switch to a known value and repeat at the second known value.

5.7.3 Insert the probe into the soil or water. Note that the tip is insulated from the probe itself and both the tip and probe must make good contact with the soil.

5.7.4 Adjust the rotating knob until a null is heard.

5.7.5 Read the resistivity directly from the scale where the rotating knob indicator is pointing. Normally the scale is calibrated in ohm-centimeters.

6.0 ANALYSIS

The determination of resistivity based on the measurements described previously is given in this section. Further analysis of soil resistivity related to corrosion and cathodic protection can be found in the literature.[7.3–7.5]

6.1 Average Resistivity

6.1.1 Option 1: Direct Current Four-Pin Measurement

6.1.1.1 Calculate the resistance from the DC test in Sections 5.1.10, 5.1.11, and 5.1.12 using Equations (6.1) and (6.2):

$$R_1 = \frac{(V_{1\ \mathrm{on}} - V_{1\ \mathrm{off}})}{I_1} \tag{6.1}$$

$$R_2 = \frac{(V_{2\ \mathrm{on}} - V_{2\ \mathrm{off}})}{I_2}, \tag{6.2}$$

where

R_1, R_2	resistance of test 1 and test 2, respectively (ohms)
$V_{1,\mathrm{on}}$, $V_{2\ \mathrm{on}}$	potential between P1 and P2, respectively, with current applied (volts)
$V_{1\ \mathrm{off}}$, $V_{2\ \mathrm{off}}$	potential between P1 and P2, respectively, with current off (volts)
I_1, I_2	current C1 and C2 (amperes)

6.1.1.2 Calculate the values for the tests at different current values in the same manner as detailed in Section 6.1.1.1.

6.1.1.3 If the resistance values are not similar, repeat the test until similar results are obtained.

6.1.1.4 The average of the resistance values is determined by Equation (6.3):

$$R = \frac{(R_1 + R_2)}{2}, \tag{6.3}$$

where

R	resistance (ohms)
R_1, R_2	calculated resistance from test 1 and test 2, respectively (ohms)

6.1.1.5 Calculate the resistivity using Equation (6.4), with the spacing in centimeters:

$$\rho = 2\pi a R, \tag{6.4}$$

where

- ρ resistivity (ohm-centimeters)
- π constant = 3.141593
- a pin spacing and average depth of measurement (centimeters)
- R resistance calculated in Equation (6.3; ohms)

6.1.1.6 If the pin spacing measurement is in feet, then Equation (6.5), adapted from Equation (6.4), can be used, but note that the answer is still in ohm-centimeters:

$$\rho = 191.5\,aR, \tag{6.5}$$

where

- ρ resistivity (ohm-centimeters)
- a pin spacing and average depth of measurement (feet)
- R resistance calculated in Equation (6.3; ohms)

6.1.1.7 Example Calculation
Given that

$$V_{1\text{ on}} = 5.2\text{ V},\ V_{2\text{ on}} = 4.9\text{ V}$$
$$V_{1\text{ off}} = 0.2\text{ V},\ V_{2\text{ off}} = -0.2\text{ V}$$
$$I_1, I_2 = 0.25\text{ A}$$
$$a = 1.5\text{ m (4.92 ft)} = 150\text{ cm},$$

from Equation (6.1),

$$R_1 = \frac{(5.2 - 0.2)\text{V}}{0.25\text{ A}} = 20.0\,\Omega;$$

from Equation (6.2),

$$R_2 = \frac{(4.9 - (-0.2))V}{0.25\text{ A}} = 20.4\,\Omega;$$

from Equation (6.3),

$$R = \frac{(20.0 + 20.4)\Omega}{2} = 20.2\,\Omega;$$

from Equation (6.4),

$$\rho = 2\pi 150\,\text{cm} \times 20.2\,\Omega = 19{,}038\,\Omega\text{-cm}.$$

6.1.2 Option 2: Alternating Current Four-Pin Measurement

6.1.2.1 Calculate the resistivity for the four-pin AC test using Equation (6.6):

$$\rho = 2\pi aR, \tag{6.6}$$

where

ρ resistivity (ohm-centimeters)

π constant = 3.141593

a pin spacing and average depth of measurement (centimeters)

R resistance measured in Section 5.2.9 (ohms)

6.1.2.2 If the pin spacing measurement is in feet, then Equation (6.7), adapted from Equation (6.6), can be used, but note that the answer is still in ohm-centimeters:

$$\rho = 191.5\,aR, \tag{6.7}$$

where

ρ resistivity (ohm-centimeters)

a pin spacing and average depth of measurement (feet)

R resistance measured in Section 5.2.9 (ohms)

6.1.2.3 Example Calculations

Given that

$$a = 1.5\,\text{m}\,(4.57\,\text{ft}) = 150\,\text{cm}$$

$$R = 18\,\Omega,$$

from Equation (6.6),

$$\rho = 2 \times 3.1416 \times 150\,\text{cm} \times 18\,\Omega = 16{,}964\,\Omega\text{-cm},$$

or from Equation (6.7),

$$\rho = 191.5 \times 4.92\,\text{ft} \times 18\,\Omega = 16,964\,\Omega\text{-cm}.$$

6.1.3 Option 3: Unequal Pin Spacing

6.1.3.1 Calculate the resistance from the DC test in Sections 5.3.11, 5.3.12, and 5.3.13 using Equations (6.8) and (6.9):

$$R_1 = \frac{(V_{1\text{ on}} - V_{1\text{ off}})}{I_1} \tag{6.8}$$

$$R_2 = \frac{(V_{2\text{ on}} - V_{2\text{ off}})}{I_2}, \tag{6.9}$$

where

R_1, R_2	resistance of test 1 and test 2, respectively (ohms)
$V_{1\text{ on}}$, $V_{2\text{ on}}$	potential between P1 and P2, respectively, with current applied (volts)
$V_{1\text{ off}}$, $V_{2\text{ off}}$	potential between P1 and P2, respectively, with current off (volts)
I_1, I_2	current C1 and C2 (amperes)

6.1.3.2 Calculate the values for the tests at different current values in the same manner as detailed in Section 6.1.3.1.

6.1.3.3 If the resistance values are not similar, repeat the test until similar results are obtained.

6.1.3.4 The average resistance values are determined by Equation (6.10):

$$R = \frac{(R_1 + R_2)}{2}, \tag{6.10}$$

where

R	resistance (ohms)
R_1, R_2	calculated resistance from test 1 and test 2, respectively (ohms)

6.1.3.5 Calculate the resistivity using Equations (6.11) through (6.13):

$$\rho = \frac{(1 - \alpha^2)}{2\alpha} \times R, \tag{6.11}$$

$$\alpha = \frac{a}{(a + 2b)}, \tag{6.12}$$

$$x = \frac{(a + 2b)}{2}, \tag{6.13}$$

where

- ρ soil resistivity (ohm-centimeters)
- a spacing between inside potential pins (P1 and P2; centimeters)
- b spacing between potential pin and current pin (P1 and C1 or P2 and C2; centimeters)

6.1.4 Option 4: Direct Current Soil Box (Four Terminals)

6.1.4.1 Average the resistance values obtained from Equation (5.4) with Equation (6.14):

$$R = \frac{(R_1 + R_2)}{2}, \tag{6.14}$$

where

- R resistance (ohms)
- R_1, R_2 calculated resistance from test 1 and test 2, respectively (ohms)

6.1.4.2 Calculate the resistivity using Equation (6.15):

$$\rho = \frac{RA}{L}, \tag{6.15}$$

where

- ρ resistivity (ohm-centimeters)
- A cross-sectional area of box between potential pins P1 and P2 (centimeters squared)
- L distance between pins P1 and P2 (centimeters)
- R resistance calculated in Equation (6.14) (ohms)

Note that if the soil box is constructed where $A = L$, as most soil boxes are made, then Equation (6.15) becomes Equation (6.16):

$$\rho = R, \tag{6.16}$$

where

- ρ resistivity (ohm-centimeters)
- R resistance from Equation (6.14) (ohms)

That is, the resistivity is equal to the resistance, but the resistivity is in ohm-centimeters and the resistance is in ohms.

6.1.4.3 As an example, if $R = 2{,}000\ \Omega$, then $\rho = 2{,}000\ \Omega$-cm.

6.1.5 Options 5 and 6: Alternating Current Soil Box (Four Terminals and Two Terminals)

6.1.5.1 Calculate the resistivity for the AC soil box resistivity for either the four-terminal or two-terminal box using Equation (6.17):

$$\rho = \frac{RA}{L}, \tag{6.17}$$

where

- ρ resistivity (ohm-centimeters)
- A cross-sectional area of box between potential pins (P1 and P2; centimeters squared)
- L distance between pins (P1 and P2; centimeters)
- R resistance measured in Section 5.4 (ohms)

Note that if the soil box is constructed where $A = L$, then Equation (6.17) becomes

$$\rho = R, \tag{6.18}$$

where

- ρ resistivity (ohm-centimeters)
- R resistance measured in Section 5.5 or 5.6 (ohms)

That is, the resistivity is equal to the resistance, but the resistivity is in ohm-centimeters and the resistance is in ohms.

6.1.5.2 As an example, if $R = 2{,}000\ \Omega$, then $\rho = 2{,}000\ \Omega$-cm.

6.1.6 Resistivity Probe

6.1.6.1 The scale reading is read directly in resistivity, normally in ohm-centimeters.

6.1.6.2 Note that the soil resistivity is a "point" measurement, that is, a measurement of the resistivity at the probe and the readings may only provide a general indication of resistivity as the null point can be broad, in which case the range of resistivity would be large.

6.2 Soil Resistivity Layer Analysis

6.2.1 One method to predict the resistivity of layers is to apply Barnes Layer analysis. This applies a parallel resistance theory to the layers. It is recognized that the accuracy is not precise, in part due to the method of averaging the resistivity in the test itself.

6.2.2 Figure 6.1 illustrates three layers of soil resistivity that are to be predicted based on the average soil resistivity measured to three different depths (a_1, a_2, and a_3).

6.2.3 Barnes layer analysis assumes that the layers are in parallel and that A_{Layer} is the thickness of each layer. The resistance must decrease with depth for this method to be valid (not necessarily resistivity, but resistance). This is because more parallel resistance is added by increasing the depth.

The parallel resistance formula is given in Equation (6.19):

$$R_t = \frac{1}{\left(\frac{1}{R_1} + \frac{1}{R_2} + \frac{1}{R_3}\right)}, \tag{6.19}$$

where

R_t	total resistance (ohms)
R_1, R_2, R_3	resistance of each parallel branch (ohms)

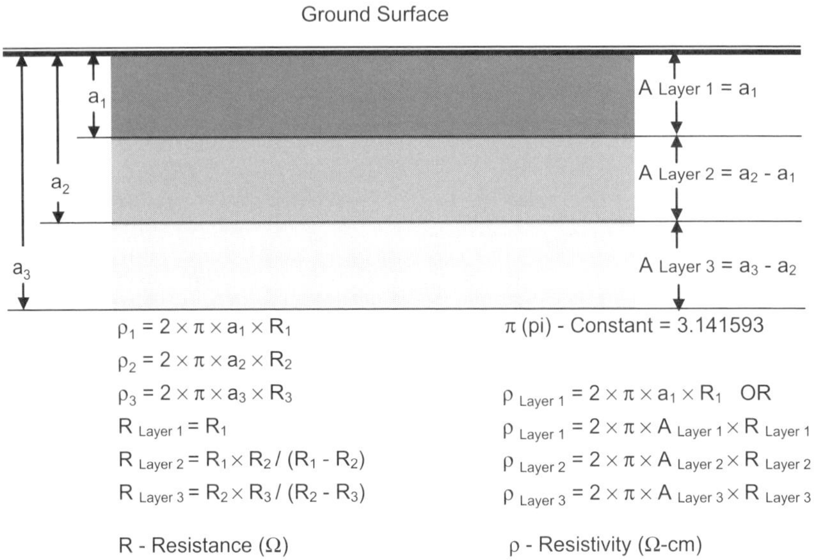

Figure 6.1 Soil resistivity layers and Barnes Layer calculations are shown.

Or if there are only two parallel resistors, one unknown resistor can be calculated using Equation (6.20) if the other resistor and the total resistance are known:

$$R_2 = \frac{R_1 R_t}{(R_1 - R_t)}. \tag{6.20}$$

In this case, R_t is the resistance value measure to depth a_2, and R_1 is the resistance value measured to depth a_1. Therefore the resistance of layer 2 ($R_{\text{Layer 2}}$) is given by Equation (6.21):

$$R_{\text{Layer 2}} = \frac{R_1 R_2}{(R_1 - R_2)}, \tag{6.21}$$

where

$R_{\text{Layer 2}}$ resistance of layer 2 (ohms)
R_1 resistance measured to depth a_1 from surface (ohms)
R_2 resistance measured to depth a_2 from surface (ohms)

The resistivity of the layer is then calculated by Equation (6.6), with new symbols applied as in Equation (6.22):

$$\rho_{\text{Layer 2}} = 2\pi\, A_{\text{Layer 2}} R_{\text{Layer 2}}, \tag{6.22}$$

where

$\rho_{\text{Layer 2}}$	resistivity of $A_{\text{Layer 2}}$ (ohm-centimeters)
π	constant = 3.141593
$A_{\text{Layer 2}}$	depth of layer 2 (centimeters; see Figure 6.1)
$R_{\text{Layer 2}}$	resistance of layer 2 calculated previously (ohms)

6.2.4 Example Calculation

Given that

$$R_1 = 18\,\Omega \quad a_1 = 150 \text{ cm}$$
$$R_2 = 7\,\Omega \quad a_2 = 300 \text{ cm}$$
$$R_3 = 4\,\Omega \quad a_3 = 450 \text{ cm}$$

$(R_{\text{Layer 1}} = R_1 = 18\,\Omega)$

- Layer 1

  Layer 1 is the same as that measured to a spacing (depth) of a_1; therefore, from Equation (6.4) or (6.6),

$$\rho_1 = 2 \times 3.141593 \times 150\,\text{cm} \times 18\,\Omega = 16{,}964\,\Omega\text{-cm}$$

$$\rho_{\text{Layer 1}} = \rho_1 = 16{,}964\,\Omega\text{-cm}.$$

- Layer 2

  From Equation (6.20),

$$R_{\text{Layer 2}} = \frac{18 \times 7}{(18 - 7)} = 11.45\,\Omega$$

$$A_{\text{Layer 2}} = 300\,\text{cm} - 150\,\text{cm} = 150\,\text{cm}.$$

From Equation (6.17),

$$\rho_{\text{Layer 2}} = 2 \times 3.141593 \times 150\,\text{cm} \times 11.45\,\Omega = 10{,}791\,\Omega\text{-cm}.$$

- Layer 3
  The third layer ($\rho_{\text{Layer 3}}$) is calculated in a similar manner to the second layer ($\rho_{\text{Layer 2}}$).
  Modifying Equation (6.20) to layer 3 gives us Equation (6.23),

$$R_{\text{Layer 3}} = \frac{R_2 R_3}{(R_2 - R_3)} = \frac{7 \times 4}{(7 - 4)} = 9.33\,\Omega \tag{6.23}$$

$$A_{\text{Layer 3}} = 450\,\text{cm} - 300\,\text{cm} = 150\,\text{cm}.$$

Modifying Equation (6.17) for layer 3,

$$\rho_{\text{Layer 3}} = 2\pi A_{\text{Layer 3}} R_{\text{Layer 3}} \tag{6.24}$$

$$\rho_{\text{Layer 3}} = 2 \times 3.141593 \times 150\,\text{cm} \times 9.33\,\Omega = 8{,}793\,\Omega\text{-cm}.$$

7.0 REFERENCES

7.1 F. Wenner, "A Method of Measuring Earth Resistivity," Bureau of Standards, Bulletin 12, 3 (1916): pp. 469–482.

7.2 L.S. Palmer, "Examples of Geoelectrical Surveys," AIEE, paper no. 2791M (Bureau of Standards, 1958), pp. 231–244.

7.3 A.W. Peabody, *Control of Pipeline Corrosion*, 2nd ed., ed. R.L. Bianchetti (Houston, TX: NACE, 2001), pp. 85–90, 106.

7.4 M.E. Parker, E.G. Peattie, *Pipeline Corrosion and Cathodic Protection* (Houston, London, Paris, Tokyo: Gulf Publishing Company, Book Division, 1984), pp. 1–15.

7.5 W. von Baeckmann, W. Schwenk, W. Prinz, *Cathodic Corrosion Protection*, 3rd ed. (Houston, TX: Gulf Publishing Company, 1997), pp. 114–118.

Index